Marine Eutrophication
A Global Perspective

W0246408

Michael Karydis and Dimitra Kitsiou
Department of Marine Sciences
University of the Aegean
Mytilene
GR81100 Greece

CRC Press
Taylor & Francis Group
Boca Raton London New York

CRC Press is an imprint of the
Taylor & Francis Group, an **informa** business

A SCIENCE PUBLISHERS BOOK

Cover picture is a 2013 oil painting on canvas by Michael Karydis. Reproduced by permission

CRC Press
Taylor & Francis Group
6000 Broken Sound Parkway NW, Suite 300
Boca Raton, FL 33487-2742

First issued in paperback 2021

© 2020 by Taylor & Francis Group, LLC
CRC Press is an imprint of Taylor & Francis Group, an Informa business

No claim to original U.S. Government works

Version Date: 20190627

ISBN-13: 978-0-367-77661-9 (pbk)
ISBN-13: 978-0-8153-6905-9 (hbk)

Library of Congress Cataloging-in-Publication Data

Names: Karydis, Michael, author. | Kitsiou, Dimitra, author.
Title: Marine eutrophication : a global perspective / Michael Karydis and
    Dimitra Kitsiou (Department of Marine Sciences, University of the Aegean,
    Mytilini, Greece).
Description: Boca Raton, FL : CRC Press, 2019. | "A science publishers book."
    | Includes bibliographical references and index.
Identifiers: LCCN 2019019384 | ISBN 9780815369059 (hardback)
Subjects: LCSH: Marine eutrophication. | Marine ecology.
Classification: LCC QH91.8.E87 K37 2019 | DDC 577.7--dc23
LC record available at https://lccn.loc.gov/2019019384

Visit the Taylor & Francis Web site at
http://www.taylorandfrancis.com

and the CRC Press Web site at
http://www.crcpress.com

Preface

The phenomenon of excessive growth of algae due to elevated nutrient loading has been known since the end of the 19th century. However, it was recognized "problem" in Scandinavian lakes at the beginning of the 20th century and later to the marine environmentalists under the term "eutrophication". During the last few decades, the problem of eutrophication has been intensified due to anthropogenic activities. This type of eutrophication is known as "cultural eutrophication". The main sources of cultural eutrophication are river outflows, sewage discharges, fertilizers' washing off and nutrient atmospheric deposition. Social, economic and environmental impacts of eutrophication, fairly often, appear in coastal waters where many activities have to be accommodated, some of them conflicting with each other. A lot of work has been carried out worldwide to understand the ecological processes under the pressure of nutrient inputs as well as to monitor causes and effects.

Marine eutrophication is, in most cases, and irrespective of the cause a transboundary problem. Luckily this problem has received the attention it deserves at political level and many states in regional seas have cooperated and signed international conventions that include measures for eutrophication control and remediation. These agreements, the outcome of the interfacing between science and policy, have opened a new era in marine environmental management. It is now widely accepted that the management of the problem requires collaboration from a wide spectrum of scientific disciplines, competent authorities, institutions as well as public participation. A new dimension to the problem which is expected to facilitate the handling of eutrophication issues is Marine Spatial Planning. This emerging legal framework is expected to eliminate conflicts between different human activities in the marine environment, set priorities in the use of marine resources and eventually minimize the impact.

The contents of this book have been designed and written having in mind the views expressed in the previous paragraph: to present eutrophication in a two-fold way: as a problem within the marine ecosystem but mainly as an issue that requires a holistic approach; this way the book is addressed to a wide readership. The contents have been organized into four sections. The first section "nature of eutrophication" presents the different aspects of the problem at ecosystemic level. It will be particularly useful to readers without prior knowledge of the subject especially if their discipline is not related to marine sciences. The second section is devoted to methodological procedures regarding data analysis. There are voluminous sets of data derived from marine eutrophication studies. It is therefore necessary to process them in order to answer questions referring to quantitative assessment of eutrophication, seasonal trends, spatial trends, future trends as well as mapping marine eutrophication in a GIS environment. In addition, as eutrophication is approached in a holistic way including social and economic aspects, an outline of multicriteria analysis is presented with applications.

The third chapter of the book is devoted to the governance of the regional seas. Governance involves regimes which are means international conventions, aggregations of policies and institutions working together to achieve good marine environmental quality. The main emphasis in this section is on policies of marine eutrophication and management practices. International conventions are presented focusing on the marine eutrophication component as well as on activities and practices at an international level. The interdisciplinary character of eutrophication issues is stressed and the need for measures and correctives is discussed.

The last chapter, chapter four, is an extensive review on water quality and ecosystems' health in the main regional seas. In every regional sea, a brief description of the physiography of the area is provided that is the boundaries of the system, bathymetry, hydrology and climate; this background information together with the main habitats and pressures, presented in the next Chapter, help the understanding of nutrient and phytoplankton dynamics, ecosystem's vulnerability and finally the impacts of eutrophication at a regional level. The environmental pressures and the general condition of the sea are also described, the eutrophication problems in the area are presented and the effectiveness of the measures for mitigating eutrophication are assessed. Obviously, the main focus of the authors is on eutrophication issues, causes, effects, policies and correctives. Views on marine eutrophication future policy and prospects are presented in the closing Section of this book.

August 2019

Michael Karydis
Dimitra Kitsiou
Mytilene

Satellite Imagery
Chlorophyll Distributions in the Regional Seas

The chlorophyll concentrations satellite data are a product of the NASA Goddard Space Flight Center, Ocean Biology Processing Group (OBPG) (https://oceancolor. gsfc.nasa.gov/). They are near-surface Level-3 annual values of the year 2017 acquired from the MODIS Aqua-sensor with a spatial resolution of 4 km. Level 3 data are derived geophysical variables that have been projected onto a well-defined spatial grid over a well-defined time period.

The classification of the chlorophyll values was based on the eutrophication scale proposed by Simboura et al. (2005), which includes a 5-level classification scale; Eutrophic (> 2.21 µg/l), Higher/Upper mesotrophic (0.6–2.21 µg/l), Lower mesotrophic (0.4–0.6 µg/l), Lower mesotrophic (0.1–0.4 µg/l) and Oligotrophic (< 0.1 µg/l) representing bad, poor, moderate, good and high water quality, respectively. Here, this scale was modified by authors and the two classes representing lower mesotrophic conditions were merged; therefore, the satellite chlorophyll values were classified to four eutrophication levels; Eutrophic—E (> 2.21 µg/l), Higher/ Upper mesotrophic—UM (0.6–2.21 µg/l), Lower mesotrophic—LM (0.1–0.6 µg/l) and Oligotrophic—O (< 0.1 µg/l).

Contents

Acronyms

Aarchus Convention	The UNECE Convention on Access to Information, Public Participation and Access to Justice in Environmental Issues
Abidjan Convention	Convention for the Protection and Development of the Marine and Coastal Environment of the West and Central African Region
ASP	Amnestic Shellfish Poisoning
ASEAN	Association of Southeast African Nations
ASMO	Agreement and Monitoring Committee
BOD	Biological Oxygen Demand
BSC	Black Sea Commission
BSEC	Black Sea Economic Cooperation
BSIMAP	Black Sea Integrated Monitoring and Assessment Programme
BSPA	Coastal Marine Baltic Sea Protected Areas
BSSSC	The Baltic Sea States Subregional Cooperation
CAP	Common Agricultural Policy
Cartagena Convention	Convention for the Protection and Development of the Marine Environment in the Wider Caribbean Sea
CEP	Sub-programme concerning Information Systems for the Management of Marine and Coastal Resources
CIESM	International Commission for the Scientific Exploration of the Mediterranean Sea
COBSEA	Coordinating Body on the Seas of East Asia
DSP	Diarrheic Shellfish Poisoning
EC	European Commission, the Administration of the European Union
EEA	European Environmental Agency
EEZ	Exclusive Economic Zones
EIA	Environmental Impact Assessment
EMS	Eastern Mediterranean Sea

EPA	Environmental Protection Agency
EQ	Environmental Quality
EU	European Union
FAO	Food and Agriculture Organization of United Nations
GESAMP	Joint Group of Experts on the Scientific Aspects of Marine Environmental Protection
GIWA	Global International Waters Affair
HABs	Harmful Algal Blooms
HELCOM	Baltic Marine Environmental Protection Committee
ICZM	Integrated Coastal Zone Management
LBS	Land Based Sources
LOS, UNCLOS or LOST	Law of the Sea Convention, United Nations Convention on the Law of the Sea, Law of the Sea Treaty
MAP	Mediterranean Action Plan
MED POL	Mediterranean Pollution Assessment and the Control Component of MAP
MSFD	Marine Strategy Framework Directive
MSP	Marine Spatial Planning
NSP	Neurotoxic Shellfish Poisoning
OECD	Organization for Economic Cooperation and Development
OSPAR	Oslo and Paris Conventions
ROMPE	Regional Organization for the Protection of the Marine Environment
PSP	Paralytic Shellfish Poisoning
RSC	Regional Seas Conventions
RSP	Regional Seas Programme
UNEP	United Nations Environment Programme
UNESCO	United Nations Educational, Scientific and Cultural Organizations
UNFAO	United Nations Food and Agriculture Organization
UWWT	Urban Waste Water Treatment Directive
WACAF	West and Central African Region
WCR	Wider Caribbean Region
WFD	Water Framework Directive
WMS	Western Mediterranean Sea
WHO	World Health Organization
WWF	World Wildlife Fund

1

The Nature of Eutrophication

1.1 Environmental Issues in the Regional Seas

The marine environment offers innumerable services and goods to mankind. In spite of the fact that the seas cover approximately 71% of Earth's surface including oceans, seas and transitional waters, only a small percentage (about 1%) of the marine resources has been explored so far. In addition to biological resources, the marine environment contains resources rich in oil, gas, nodules of manganese, copper, nickel and many more valuable minerals. However, the most striking fact is that half of the world's population lives by the sea or near the seashore: population densities as high as 10,000 habitats per square kilometer (Small and Nicholls, 2003) is common along coastal areas. A great part of the infrastructure, ports, oil refineries, power stations, shipyards, terrestrial roadways and various types of factories have been built on the coastal zone. All these activities exercise environmental pressures on the marine environment; these pressures can be local affecting a narrow strip of the coastal marine area but as there are no boundaries in the marine environment, pollutants can be transferred by currents and spread over a wide area causing what is known as "transboundary pollution". A form of transboundary pollution is also airborne, a mechanism that had been underestimated in the past (UNEP, 2003b). Most of the pollution problems originate from land based sources. These types of discharges that reach the sea through rivers, coastal disposals, canals, underwater pipelines or through run-offs, include organic compounds, fertilizers, industrial effluents, pesticides and insecticides.

Marine pollution was recognized as a threat for the marine environment in semi-enclosed seas characterized by limited communication with the ocean; two examples are the Mediterranean Sea and the Baltic Sea. When it is finally accepted that an environmental problem does exist at a wide scale, then the problem is "politized" and the member states surrounding the regional sea as a rule come to an agreement which gets the form of an International Convention. The two Conventions referring to our paradigm are the Barcelona Convention, signed in 1976 for the protection of the

Mediterranean Marine Environment and the Helsinki Convention for the protection of the Baltic Sea signed in 1974. In other words, it was realized that an international cooperation among states, at least at the level of every regional sea was needed. According to Benedict (1986): "*Somehow leaders and government processes and budget makers must accustom themselves to a new way of thinking. The scientific issues are complex, interpreting many different disciplines, often at the frontier of discovery*".

The legal framework regarding marine environmental protection and management has been expanding over many regional seas around the world (DiMento and Hickman, 2012) for an additional reason: it is only over the last twenty years or so that the importance of Ecosystem Services (ES) has been appreciated not only by scientists working in the marine environment but also by policy makers (Turner and Schaafsma, 2015). All these ecosystem services are direct or indirect resources necessary for good ecosystems' health and human prosperity. Ecosystem services are categorized into two groups: the first group refers to ecosystem functions which include ecosystem integrity, ecosystem health and biodiversity conservation. The second group refers to ecosystem services that offer an immediate profit to humans such as exploitation of fishery stocks and freshwater production from desalination, fish food through aquaculture and biotechnology products such as medicines. This group of services is also known as ecosystem benefits (Daily, 1997; Ruhl et al., 2007; Karydis et al., 2015).

The marine environment is also used for sewage disposal. The organic load is decomposed by marine organisms, mainly bacteria into carbon dioxide and water; nitrogen and phosphorus are released as inorganic nitrogen and phosphorus and assimilated by the phytoplankton. This process can also be considered as an ecosystem service, however, excessive sewage discharges cause diverse effects on the marine ecosystem and seawater quality, a phenomenon that has been identified since the 19th century. This fact led scientists and engineers to construct underwater sewers for dispersion of urban waste. Unfortunately, the dogma "*the solution to pollution is dilution*" did not quite work when the carrying capacity of a system is exceeded. Ketchum (1969, 1983) stressed the importance of the carrying capacity of the marine environment to dilute, disperse and decompose the wastes of "*an active, vigorous and effluent society*". The same author stressed that there were limits to the capacity of marine ecosystems to assimilate sewage material and in many cases this capacity had already been exceeded. The first International Conference on "*Waste Disposal in the Marine Environment*" was held in 1959 in Berkeley, focusing on ways to mitigate nutrient pollution and excessive growth of algae. In addition to eutrophication due to sewage disposal, fertilizers have contributed to the problem of marine eutrophication. These days, marine eutrophication is the first or second form of pollution in many regional seas around the world (DiMento and Hickman, 2012).

Although the first stages of marine eutrophication policy were mainly focusing on environmental protection, it was progressively accepted that in addition to scientific issues, social and economic aspects should also be taken into account. According to UNEP (Haas, 1990), environmental problems should not be considered as negative externalities; they must be examined in connection with a wide spectrum of interacting stresses in an integrative way. It is therefore obvious that policies that need to be applied should be characterized by "*comprehensive*" decisions rather than a "*piecemeal outlook and vision*". Schulman (1975) urged that "*these policies are characterized by an indivisibility in the political commitment and resources they require for success*". Environmental problems including eutrophication cannot be dealt with in isolation. Any kind of policy against pollution requires consideration of the sources of pollution (Cloern, 2001). Pollution sources, include human living conditions since specific habits have a negative effect on environmental quality. These habits are connected with demographic, social and cultural forces exercised in overcrowded areas. This coordinated policy can be materialized through environmental management taking into account marine resources, economic growth, social trends and habits as well as marine ecosystems' integrity. However, there are limits regarding the capacity of management practices.

Due to the size of megacities and their impact on the coastal environment, big industrial infrastructures along the coastal zone and enormous offshore installations for oil and natural gas production, the idea of Marine Spatial Planning (MSP) "*has experienced an intense and dynamic growth at an international scale in recent years and several practices have emerged from different continents and countries*" (Kyvelou and Pothitaki, 2017). As Marine Spatial Planning is defined as "*a process of analyzing and allocating parts of three dimensional marine spaces (or ecosystems) to specific uses and objectives, to achieve ecological, economic and social objectives that are usually specified through a political process*" (UNESCO, 2013), it is obvious that the eutrophication component forms a substantial part of the whole planning.

1.2 Objectives and Structure of the Study

Aspects and thoughts mentioned above specify the framework and boundaries of the contents of this volume. Eutrophication will be examined only as an environmental problem in conjunction with the pressures, the legal framework and the practices that have been adopted so far. In the first section, the nature of marine eutrophication is presented; this Chapter deals with definitions of eutrophication, natural cause-effect relationships, eutrophication scaling, toxic algal blooms and remedies. The second Chapter is devoted to data analysis methods. The sets of variables related to eutrophication are voluminous and therefore data processing is necessary to come to conclusions regarding ecosystem quality, marine water quality, spatial and temporal trends of eutrophication as well as future trends. The general idea of each data processing method is presented, followed wherever possible, by a discussion on the methodological limitations and shortcomings. The readers can find in the proposed literature technical details on the methods as well as relevant applications. The third section presents the international legal framework and any additional tools for the governance of the regional seas. The fourth Chapter includes an assessment of water quality and ecosystems' health of the major regional seas, mainly from the point of view of eutrophication. The authors have attempted a rather holistic approach trying to understand eutrophication problems in connection with coastal management practices and the emerging field of Marine Spatial Planning.

1.3 Marine Eutrophication: Historical Background and Definitions

It is now well known that excessive discharges in the marine environment triggers excessive production of plant material, namely phytoplankton and phytobenthos. This phenomenon known as eutrophication is among the serious threats in many regional seas around the world (Karydis and Kitsiou, 2014; Karydis, 2017). The history of eutrophication goes back to the beginning of the 20th century when it was identified and described for the first time in the lakes down in the lowlands around the Baltic Sea (Hutchinson, 1967). The outstanding botanist and limnologist Einar Naumann, working on lake ecology realized observing the different trophic state among lakes in the Baltic that "the differences expressed themselves largely in terms of a greater mass of phytoplankton giving distinct planktogenic color to the water and in quiet summer weather producing the appearance of dense superficial accumulation of phytoplankton, usually spoken in the contemporary English as water blooms was what he named eutrophication" (Hutchinson, 1967). On the contrary, such phenomena were not observed in mountain areas. It was believed that waters draining from Paleozoic rocks of the Scandinavian mountains were deficient in nitrogen and phosphorus compounds whereas, waters entering the lakes after bleaching the cultivated lowlands were much richer in plant nutrients. The situation characterizing poor phytoplankton growth in mountainous lakes was assigned as "oligotrophy" whereas the conditions characterized by excessive plant growth were assigned as "eutrophic". Historically it is Naumann who coined the terms "oligotrophy" and "eutrophication" referring to lake trophic conditions (Hutchinson, 1967). Carlson (1977) proposed a scale on "trophic classification" for lakes. Carlson categorized the trophic state of lakes into three main classes: oligotrophic, mesotrophic and eutrophic. He also added two additional classes describing extreme conditions, beyond both ends of the proposed scale: the extreme oligotrophic state was defined as "hyperoligotrophy" whereas the extreme eutrophic as "hypertrophy". The main classification criteria were chlorophyll and phosphorus concentrations as well as Secchi depth (SD in meters). The term eutrophic originates from the Greek word eutrophos (εύτροφος) meaning "well nourished"; the prefix eu means good and "trophein" means "to nourish" according to encyclopedia Britannica. The term eutrophication was redefined in a broader sense referring to "any and all nutritive substances" (Halser, 1947). All these definitions of eutrophication have been proposed to describe nutrient enrichment in lakes.

Meanwhile eutrophication was identified as a problem in the marine environment, especially in coastal areas. In addition, marine ecologists started to understand secondary effects due to eutrophic conditions. There was, therefore the need to revise the definition. Steele (1974) proposed that "*Eutrophication is the increase of growth rate of the algae, following a faster rate of nutrient increase in the marine environment as well as the consequences*". A more comprehensive definition on eutrophication mentioning a number of secondary effects was given by Vollenweider (1992): "*Eutrophication—in its more generic definition that implies to both fresh and marine waters—is the process of enrichment of waters with plant nutrients, primarily nitrogen and phosphorus that stimulates aquatic primary production and its more manifestations, lead to visible algal blooms, algal scum, enhanced benthic algal growth of submerged and floating macrophytes"*. A similar and very descriptive definition also referring to the effects of eutrophication has been proposed by Carbiener in a French Official Report (Carbiener, 1992): *"Eutrophication is a process in which the bioavailability of nutrients in the considered recipient is increased. It becomes a nuisance if the concentrations in nutrients exceeds some threshold values that vary in a large range according to the typology of the ecosystem. This state of nuisance results in lack of diversity and complexity of the considered ecosystem, involving perturbation (if not disappearance) of the secondary productivity level. Eutrophication may be linked to and be part of both organic and biological pollution on the one hand and cause toxic effects on the other hand"*.

Eutrophication is a complex ecosystemic procedure and therefore there are different points of view that can be the focal point of the definition. Gray (1992) proposed a definition for eutrophication exempting the effects of toxic compounds entering the marine ecosystem. According to Gray *"Eutrophication occurs when nutrients are added to the body of water they load provided they are not toxic compounds and provided that there is sufficient light to increase autotroph growth and also to increase heterotroph growth"*. A much shorter definition has been suggested by Nixon (1995): *"Eutrophication is an increase in the rate of supply of organic matter to an ecosystem"*. Nixon's definition is not taking into account that nitrogen and phosphorus, the two limiting nutrients in primary productivity can be recycled many times contributing to carbon fixation. In addition, monitoring primary productivity is a rather costly method and therefore data availability will be limited (Nixon, 2009). However, Nixon's definition was appealing to marine scientists as Nixon corrected his definition emphasizing that "eutrophication is a process rather and not a state" (Ferreira et al., 2011), although from the management point of view it leaves a degree of uncertainty for interpretation in court cases. This definition seems to cover a specific aspect of eutrophication which is linking the responses of marine water masses to organic industrial and urban effluents. Another definition (Heip, 1995) focuses on the benthic community mentioning that "nutrient increase leads to a number of problems when benthos is considered". The effects of nutrient loading on primary productivity may not always be direct: this can be due to the fact that nutrients are transferred from long distances as living biomass or as dead organic matter.

Due to the seriousness of the problem, concerns about eutrophication resulted in political action implemented by national legislation and international conventions. Definitions of eutrophication can therefore be found in technical reports by UNEP, European Union Directives as well as international conventions such as OSPAR, HELCOM, the Barcelona Convention for the protection of the marine environment. OSPAR has provided a rather comprehensive definition of eutrophication (OSPAR, 1998): *"Eutrophication means the enrichment of water by nutrients causing an accelerated growth of algae and higher forms of plant life to produce an undesirable disturbance to the balance of organisms present in the water and to the quality of the water concerned, and therefore refers to the undesirable effects resulting from anthropogenic enrichment by nutrients ..."*. Definitions of eutrophication can also be found in national legislation in USA (Clean Water Act PL 92-500 (1972) and the Coastal Zone Management Act PL 92-583, 1972). The setting up of legal frameworks led to a need for a legal agreement on definitions. In a report on eutrophication in Europe in coastal waters published by the European Environmental Agency (EEA, 1999), although they agree with Nixon's definition, they make it clear, possibly for legal reasons that "in this report eutrophication means enhanced primary production due to excess supply of nutrients from human activities, independent of the natural productivity level for an area in question". However, the European Union Directive concerning Urban Waste Water Treatment (1991a), focuses on enrichments by nitrogen and phosphorus because it is these two nutrients that have to be removed during sewage treatment: *"The enrichment of water by nutrients, especially compounds of nitrogen and/or phosphorus, causing accelerated growth of algae and higher forms of plant life to produce an undesirable disturbance to the water balance of organisms present in water and to the quality of the water concerned"*. This definition was also accepted by·a working group on eutrophication in European ecosystems (EEA, 1999). Later UNEP (2003) proposed a definition on eutrophication with emphasis on the supply of organic matter: *"Eutrophication is defined as an environmental disturbance caused by excessive supply of organic matter"*. All these definitions show a common characteristic: they refer to eutrophication as a "disturbance". This means that the situation of the ecosystem is reversible as opposed to certain forms of pollution causing permanent environmental degradation. The definitions of eutrophication and their implications on monitoring strategies have also been reviewed by Andersen et al. (2006a).

The European Commission (EC) in 2009, created task groups within the European Marine Strategy Framework (EC, 2008) with the mandate to propose quality descriptors that could form the background for evaluating ecosystem quality. One of these fields of interest was referring to the assessment of eutrophication (Ferreira et al., 2011). It was realized that an operational definition of eutrophication was necessary for both, the evaluation of available methods and its appropriateness in a court of law. The following points were taken into account by the group of experts on eutrophication: (a) any current developments by the scientific community regarding the understanding of eutrophication processes and impacts should be built-in within the definition. In particular, it must be taken into account that the phenomenon is triggered by an excess of nutrients mainly nitrogen and phosphorus as well as silicon. This nutrient enrichment of the marine environment is to a great extent of terrestrial origin and modifies the "pristine" conditions in the marine environment (b) it is important in large marine areas to consider the spatial variability (Kitsiou and Karydis, 2017) and therefore to realize that focal points of eutrophication effects differ, depending on physical conditions including topography. The effects from the loss of seagrasses (Submerged Aquatic Vegetation—SAV) are rather serious in the shallow areas of the Danish Straits as well as parts of the Mediterranean Sea. On the contrary, in deeper systems one of the principal impacts is that of species shifts leading to harmful algal blooms (HABs). There is also a need to distinguish between naturally occurring algal blooms and discharges causing bloom episodes (Anderson and Garrison, 1997; Barale et al., 2008; Siokou-Frangou et al., 2010; Kitsiou and Karydis, 2011) and (c) as nutrient enrichment occurs naturally, it is important to assess the extent to which nutrients of anthropogenic origin increase primary productivity as well as cause possible changes in the N:P:Si stoichiometry causing a shift from diatoms to non-siliceous algae down to the level of cyanobacteria. Those shifts although not necessarily harmful, can cause an "undesirable disturbance", eventually affecting ecosystem structure and function including the provision of ecosystem goods and services (Everard, 2017). However, it has been pointed out (Ferreira et al., 2011) that "such effects do not always result from nutrient enrichment and may be triggered by other causes, including climate change, the removal of top predators by fishing, enrichment by allochthonous organic matter and contamination by harmful substances". The working group taking into account all aspects described above, came to an agreement of the following definition: "*Eutrophication is a process driven by enrichment of water by nutrients, especially compounds of nitrogen and/or phosphorus, leading to: increased growth, primary production and biomass of algae; changes in the balance of organisms; and water quality degradation. The consequences of eutrophication are undesirable if they appreciably degrade ecosystem health and/or sustainable provision of goods and services*".

It is obvious from the definitions of eutrophication in the previous paragraphs that two schools of thought have been developed: the definitions emphasizing the causal relationship between the main nutrients (nitrogen and phosphorus) and the definitions that focus on the production of organic matter. Since both nitrogen and phosphorus are considered as the limiting nutrients regulating carbon fixation and therefore plant biomass production, they seem to be widely accepted. This type of definition is being used on a scientific basis (Steele, 1974; Gray, 1992; Vollenweider, 1992; O'Sullivan, 1995; Smith et al., 1999; de Jonge et al., 2002; Kitsiou and Karydis, 2011; Karydis and Kitsiou, 2014). In addition to the wide popularity within the academic community, definitions based on the causal of nutrient–phytoplankton growth relationship, also appear in a number of legal documents such as the European Union Directive on the "*Protection of waters against pollution caused by nitrates from agricultural sources*" (EC, 1991b), by technical reports on the Mediterranean Action Plan on "Municipal Waste Water Treatment Plants in the Mediterranean Sea" (MAP, 2008) and Blue Plan launched by UNEP (UNEP, 2009).

1.4 Eutrophication Processes in the Marine Environment

It has already been mentioned that marine eutrophication can be the outcome of natural processes. These processes include coastal upwelling, sediment resuspension due to hydrodynamic conditions but mainly nutrient enrichment from atmospheric processes (UNEP, 2003b). River water even in "pristine" condition, contains nitrogen and phosphorus compounds that end into the marine environment either as soluble nutrients or as sediment particles. This form of eutrophication is also known as "autochthonous" eutrophication as opposed to eutrophication caused by human activities known as "cultural eutrophication" (Richardson and Jorgensen, 1996). Coastal eutrophication is mainly caused by land-based sources: these are agriculture, aquaculture, wastewater from treatment facilities, urban runoff and burning of fossil fuels (NRC, 2000). Cultural eutrophication has been addressed as a serious threat since the 80s (Rosenberg, 1985; Gray, 1992; NRC, 2000; Karydis and Kitsiou, 2012; Karydis and Kitsiou, 2014). Nixon (1990) had stated that eutrophication is a worldwide problem. Starting from the Baltic and the North Sea, is spreading along the rim of the Mediterranean Sea, extending further along the coastal areas of the North and South America, Africa, India, South-East Asia, Australia, China and Japan. Nixon considered population increase in coastal zones, the intensification of agricultural practices with excessive use of fertilizers, deforestation and the release of nitrogen oxides as the principal causes. Although most definitions of eutrophication emphasize on the growth of plant material, eutrophication is a multistage environmental disturbance. The initial effects induce further problems in the ecosystems' components, other than the phytoplanktonic and phytobenthic system. The most realistic model of marine eutrophication has been proposed by Gray (1992). According to Gray there are five levels of ecosystem disturbance (Figure 1.1) ranging from the initial effects to the ultimate effects depending on the quantity of nutrient loading from external sources. External nutrient supply increases growth rates of macroalgae. It was found in Nova Scotia, as early as 1977, that the excessive growth of *Laminaria longicuris* was the result of the response of the alga to increased nutrient levels (Chapman and Craigie, 1977). Later, the dramatic increase of primary production in the Dutch Wadden Sea was closely correlated to nutrient discharges from the river Rhine. Nitrogen and phosphorus addition to salt marshes and seagrass may have increased their growth rates. This supposition is supported by the observations that the seagrasses *Thalassia* and *Zostera* are nitrogen limited (Patriquin and Knowles, 1972; McRoy and Goering, 1974). Phytoplankton production increases and in some cases, the biomass concentrations get very high.

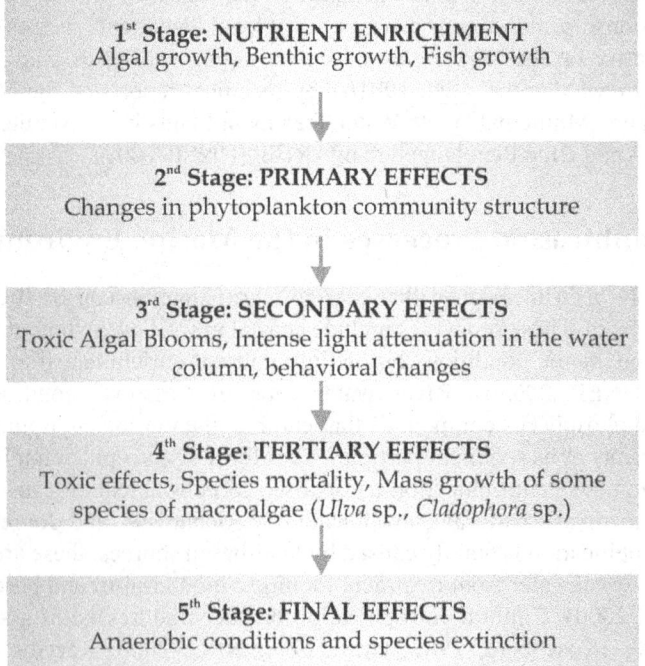

Figure 1.1 The different stages of marine eutrophication as nutrient inputs increase, proposed by Gray (1992).

This is the case with the Adriatic Sea in the Mediterranean; chlorophyll concentrations 1.2 mg.m^{-3} (Pettine et al., 2007) and phytoplankton abundance as high as 7.0×10^4 cells l^{-1} are often recorded (Vilicic et al., 2011; Vollenweider, 1992). Eutrophic conditions regarding phytoplankton growth have also been reported in the Black Sea (Karydis and Kitsiou, 2014) and the Baltic Sea (Hakanson and Bryhn, 2008; Gray, 1992). High phytoplankton biomass and high productivity rates change the population dynamics between species, a phenomenon that induces changes in the species composition of the phytoplankton community. This has been characterized by Gray (1992) as the "initial effect". Many years ago, Dugdale (1967), working on phytoplankton presented a model on nitrate limitation based on the uptake model introduced by Monod (1942). He found that the substrate constant K_s at which $V_{N4} = V_{max}/2$, where V_{max} is the maximum uptake velocity of phytoplankton species under the experimental conditions, varies among the species. This fact seems to be the key for the ecological interpretation of species competition and succession (Figure 1.2). Dugdale's model indicates that "a single species of phytoplankton or a group of species with similar kinetic characteristics dominates the population". Consequently, the growth of phytoplankton species with a small K_s is favored at low nitrate concentrations whereas species characterized by higher K_s values for nitrate uptake are favored at high nitrate concentrations. This model therefore explains why an increase in nutrient concentration changes the species population dynamics and triggers species succession in phytoplanktonic communities. Early experiments in

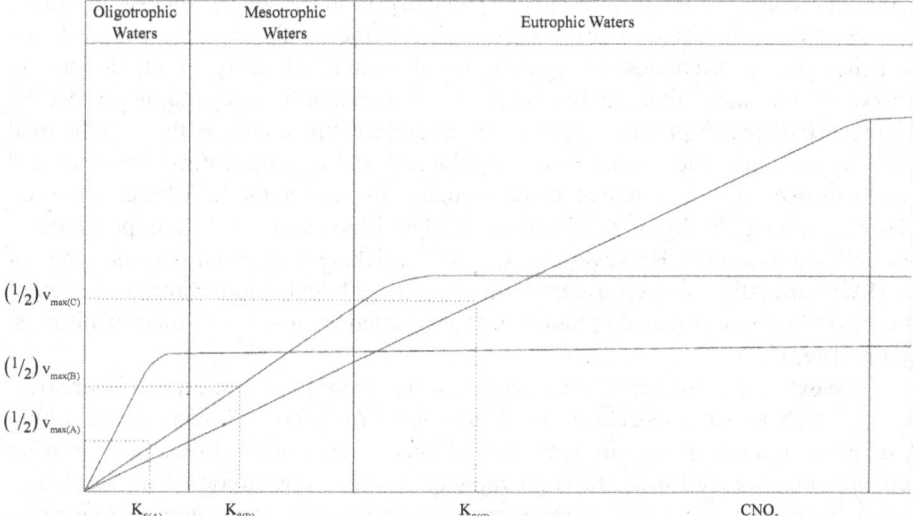

Figure 1.2 A hypothetical graph showing nitrate uptake by three microalgae (A), (B) and (C) as a function of nitrate concentrations according to Michaelis-Menten model proposed by Dugdale (1967) to elucidate nitrate kinetics by marine phytoplankton species. Low K_s values for species (A) favor its growth under oligotrophic conditions. Mesotrophic conditions favor the growth of species (B) as $K_{sb} > K_{sa}$ whereas eutrophic conditions favor the growth of species (C) over the species (A) and (B). This nutrient uptake model explains competition and succession. It is obvious that changes in species composition in marine phytoplankton communities can occur during nutrient enrichments, leading the system from oligotrophic to eutrophic (the hypothetical graph was drawn by M. Karydis).

mesocosm systems have shown that nutrient enrichment by sewage was linked to a shift in species composition from diatoms to flagellates (Taslakian and Hardy, 1976). A similar trend has been reported along the Dutch coastal area of the North Sea and has been attributed to increased nutrient input (Bennekom et al., 1975) as well as in the Baltic Sea (Graneli et al., 1990). Organic enrichment also causes changes in the benthic community (Pearson and Rosenberg, 1978; Gray, 1982; Gray et al., 1990; Gray, 1992). The interaction between phytoplankton and the benthic community has been shown in mesocosm experiments (Doering et al., 1990): the structure of the phytoplankton community under investigation was altered due to selective grazing by the benthic species. Changes in species composition can also occur in the fish community. It was found in the Baltic (Hansson and Rudstam, 1990) that the decrease in whitefish was linked with an increase in pikeperch.

Secondary effects are more complex. Reduced transparency can decrease the depth of occurrence of phytobenthic species. This symptom was observed with the macroalga *Fucus vesiculosus* in the Baltic when the spatial distribution of the macroalga was compared with historical data collected during the period 1942–1943 (Kautsky et al., 1986). In addition to reduced transparency, settling of epiphytes on the thallus of *Fucus* also inhibited light availability to the macroalga.

Nutrient experiments that trigger the growth of phytoplankton, increase the frequency of appearance of nuisance algal blooms; this fact has been noticed worldwide (Smayda, 1990; GESAMP, 1990; Hallegraeff et al., 2003; Ignatiades et al., 2007). A relevant phenomenon is the Harmful Algal Blooms (HABs); Hallegraeff (2003) has classified HABs into three classes: (a) Harmless species that simply cause discoloration in the marine environment. In extreme conditions they can be lethal to fishes and invertebrates through oxygen depletion. These types of blooms are known as "red tides" although their color can be green or blue depending on species' pigments (b) Species producing potent toxins that can find their way through the food chain to humans. These can cause neurological and gastrointestinal illnesses and (c) Nontoxic species that can cause damage in aquacultures, killing fishes by clogging their gills. More information on algal blooms and HABs is provided in the following section. However, there is still a debate linking nutrient enrichment to HABs; investigations on a Chrysochromulina polylepis bloom in the Skagerrak during 1988 showed that this bloom was supported by a normal stock of nitrogen (Gray, 1992).

The extreme effects are characterized by excessive growth of certain macroalgal species such as *Ulva* (Sfriso et al., 1988) and *Cladophora* (Baden et al., 1990). Excessive amounts of organic matter in the marine environment require significant amounts of dissolved oxygen. High rates of oxygen consumption lead to oxygen depletion (hypoxia) and at the extreme stage to absence of oxygen (anoxia), a situation that is usually followed by formation and release of hydrogen sulphide. Mortality of benthic species occurs at the phase of hypoxia. Organisms found in sand bottoms like the bivalve *Abra* and the echinoderm *Echinodesmium* show sensitivity in reduced oxygen concentrations (Pearson and Rosenberg, 1978; Baden et al., 1990).

The ultimate stage of eutrophication is characterized by anoxia, that is the conditions are anaerobic leading to death of all the planktonic and benthic species. Anaerobic bacteria dominate producing methane (CH_4) and hydrogen sulphide (H_2S).

1.5 Eutrophication Scaling

The quantification of eutrophication levels and the characterization of marine water masses into oligotrophic, mesotrophic and eutrophic water types (Ignatiades et al., 1992) has been one of the main objectives since the 1970s. Many classification schemes have been proposed so far, but most of them are based on cause and effect variables namely nutrient concentrations, chlorophyll concentrations, productivity values or indices. In addition, bioindicator species used in coastal studies (cyanobacteria and seaweeds) and environmental indicators used for eutrophication assessment on a quantitative basis have also been proposed (Karydis, 2009). There are many difficulties in classifying water masses due to seasonal variations of nutrient and chlorophyll concentrations, to species succession and to the fact that the boundaries between water types from the trophic point of view do not actually exist in the natural environment; however, a classification scheme is particularly useful because it facilitates communication among scientists as well as between scientists and policy makers, an interface which is not always "trouble-free". In addition to the scientific and administrative regime, the public participation regarding environmental issues seeks for objective criteria (Hakanson, 2006; Kitsiou and Karydis, 2011). The main reasons in classification schemes regarding eutrophication are: the difficulty to define background concentrations as reference values that are based on nutrient and chlα concentrations. Pristine states, that is, unimpacted marine areas which are defined as "*the insignificant impact of pressures on ecosystem functioning and thus approximation of the natural environment* (Thake et al., 2010). Pristine waters cannot be easily located these days and it is unrealistic to speak about pristine waters in some areas for example in some extended marine areas along the Northern and Western European coastal waters, marine areas in the U.S.A. and South Eastern Asian Coastal Waters. The problem of "pristine conditions" often arises at the onset of research on marine water quality assessment. This type of research is usually financed after the phase of "problem recognition" (Nielsen et al., 2007). There is a number of problems regarding these studies: (a) historical data are irregularly selected and therefore reliable time series of data do not go back very far, making any trend detection difficult (b) as nutrients and phytoplankton are system's variables, partitioning of nutrient and chlorophyll concentrations into natural and human origin is particularly difficult and (c) seasonality fluctuations (already mentioned) (d) the situation can be further complicated in shallow areas because of indigenous eutrophication: local topography in combination with water circulation and sediment composition may alleviate sediment resuspension. This way eutrophic trends are observed although they are not the result of anthropogenic activities. The shortcomings just mentioned indicate that there is no eutrophication scale that can be appropriate for any kind of system.

Ecological indices have been proposed since the 60s as estimators of eutrophic conditions (Wilhm and Dorris, 1968). This practice was followed by empirical indicators based on variables related to eutrophication. Ferris and Humphrey (1999) have given a definition of indicators: "as indicator may be defined a characteristic that when measured repeatedly, demonstrates ecological trends and a measure of current state quality in an area". An indicator should be characterized by a number of criteria (a) to be based on variables used in routine work (b) to be robust in seasonal fluctuations (c) to range within specific limits and (d) to define clearly oligotrophic, mesotrophic and eutrophic conditions. A number of indicators are presented under the data analysis section. There are some paper reviews on indicators for eutrophication studies by Washington (1984), Karydis (2009), Karydis and Kitsiou (2011) and Ferreira et al. (2011).

Water transparency is also a measure of eutrophic conditions. Secchi disk depth more than 11 m indicates oligotrophic conditions whereas, Secchi depths below 2 m indicate eutrophic status. A scale based on chlα concentrations had been proposed by Giovanardi and Tromellini as early as 1992 (Giovanardi and Tromellini, 1992). Michelakaki and Kitsiou (2005) have also proposed an index of eutrophication for the Mediterranean Sea: the waters were classified into oligotrophic, lower-mesotrophic, upper-mesotrophic and eutrophic. ICES has proposed chlα concentrations as a useful indicator for assessing eutrophic levels (OSPAR, 2005c). The importance of chlα concentration as an indicator for assessing eutrophic levels has been indicated by the Task Group 5 working within the framework of the Marine Strategy Framework Directive (Ferreira et al., 2011).

Background values of nutrient and chlα concentrations differ in different seas due to topography, circulation, stratification, river contributions and communication with the ocean. A good example is the comparison of eutrophic states between the Baltic Sea and the Mediterranean Sea. Eutrophication assessment in the Baltic Sea showed high scaling of chlα concentrations (Wasmund et al., 2001; Hakanson, 2006). Similar scaling proposed for the Mediterranean Sea (Ignatiades et al., 2005) showed that only 10–25% of the chlorophyll concentrations of the Baltic Sea were characterizing the same level of eutrophication.

1.6 Nutrient Limitation

Nutrient limitation is a state in the marine environment characterized by very low nutrient concentrations. It is the opposite extreme to eutrophication, very important however to understand ecosystemic processes and phytoplankton dynamics. Nutrient limitation has been studied rather extensively during the 60s and 70s under laboratory conditions using microalgae (Rhee, 1978; Goldman et al., 1979; Fogg and Thake, 1987). Most of our understanding of phytoplankton growth under nutrient limited conditions comes from laboratory work using batch and chemostat cultures. Extremely low nutrient concentrations in the culture medium, hence the marine environment, affect the cellular composition of phytoplanktonic cells, altering growth rates and buoyancy (Goldman et al., 1979) as well as resilience to toxic compounds (Karydis, 1981). Unialgal systems cannot however provide information at ecosystem's level and it is necessary to employ other experimental systems, namely, marine microcosms (see Section 2.12) or even mesocosms in order to understand nutrient limitation at a higher level of organization. Nutrient limitation *in situ* has been studied in oligotrophic environments like the Eastern Mediterranean Sea (Krom et al., 1991; Ignatiades, 1998; Ignatiades et al., 1995). Studies on nutrient limitation have been carried out at various levels: nutrient dynamics (Souvermezoglou et al., 1999; Krom et al., 2010), at population level (Goldman et al., 1979) and at the level of primary productivity (Ignatiades, 2005). Many disagreements found in literature between nutrient limitation and eutrophication processes (causes–effects) have been attributed (Richardson and Jorgensen, 1996) to the fact that these studies have been performed at different hierarchical levels and different physico-chemical conditions within the ecosystem.

1.7 The N/P Ratio

It had been proposed for the first time by Ryther and Dunstan (1971) that the average concentrations of dissolved phosphorus and nitrogen in the marine environment are almost the same as the average concentrations of these substances in plankton. This phenomenon is known as the Redfield Ratio (Redfield, 1963); any changes on the 16:1 N:P ratio was attributed to uptake processes by phytoplankton. The importance of the N:P ratio has been stressed by Ryther and Dunstan (1971) who supported the view that the observed relationship "*may be important in regulating the level of balance of nutrients in the ocean as a whole and over geological time. It is certainly not effective locally of the short time*". These authors concluded that "*it has become increasingly clear that the concept of a fixed nitrogen to phosphorus (N/P) ratio of approximately 15 to 1, either in the plankton or in the water in which it has grown, has little if any validity*". Although an N/P ratio value of about 16 is considered as ideal, such a ratio value rather rarely exists in nature. According to Ryther and Dunstan (1971), the usual values in the sea range between 5 and 15; a ratio of 10 has been characterized as a "*reasonable working value*". Similar work in Saronikos Gulf (Mediterranean Sea) based on a large number of samples (Karydis et al., 1983) showed that the N/P values calculated as annual integral means at each station varied between 19.34 and 10.12.

A 15:1 atomic ratio may be realistic for the ocean as a whole but not for the euphotic zone which accounts only 2% of the total ocean volume. Ryther and Duncan (1971) also found that in surface waters nitrogen compounds were, in most cases, depleted faster than phosphates, concluding that the surface sea waters were nitrogen limited. This general view is in contradiction with the views of Krom et al. (1991) who found that primary productivity in the Eastern Mediterranean was phosphorus limited. This view is in agreement with published work concerning other areas in the Mediterranean: Chiaudani et al. (1980) and Marchetti (1985) found that phosphorus was the limiting element in the Adriatic Sea. Mingazzini et al. (1992) also found phosphorus as a limiting factor; this is an indication that there is no conclusive evidence regarding the limiting nutrient in the Mediterranean Sea.

Decision on the limiting nutrient has been a cause of debates in U.S.A. towards banning the use of phosphate in detergents in order to mitigate eutrophication. As the aspect dominated during the 60s was that nitrogen was the limiting factor, Clark (1971) maintained in his paper "*Futility of Phosphate Detergent Ban*" that "*authorities have chosen to control the discharge of phosphorus rather than nitrogen because its source is more distinct, being almost entirely the detergent in the sewage and because it can be removed at the treatment plants easily and cheaply".*

In addition to nutrients of sewage origin, atmospheric deposition of nitrogen also has to be taken into account in budgets of anthropogenic nutrients. Many authors have supported the view that altered N/P ratios may also favor blooms of flagellate species which reduce considerably the percentage of siliceous diatoms during spring and autumn blooms (Officer and Ryther, 1980; Ryther and Dunstan, 1971; Smayda, 1990).

1.8 Eutrophication and Toxic Algal Blooms

There are cases where proliferation of microalgae is very intensive leading to exceptionally high cell numbers, possibly as high as a few million cells per liter. These dense concentrations of phytoplankton cells are known as "algal blooms" or "Red Tides". Dense microalgal concentrations are formed even by non-toxic bloom forming species, generate anoxic conditions that cause death to both fish and invertebrates. It is not known exactly how an algal bloom phenomenon breaks out but there is a consensus among marine scientists that when a delicate balance between many physical, chemical and biological factors including spatial heterogeneity and seasonal variation is disturbed beyond certain limits, a bloom formation is triggered. Sournia et al. (1991) indicated that out of the 5,000 living marine species, there are 300 species that occur from time to time at such a high number that discolor the sea surface, hence the term "red tides". Eighty out of those species have the capacity to produce toxins that affect fish, shellfish and following the food chain to humans (Hallegraeff et al., 2003). These phenomena were known since ancient times. There is a written reference (1000 B.C.) in the book of exodus (7, 20-1) of the Bible: *"...all the waters that were in the river were turned to blood. And the fish that was in the river died; and the river stank, and the Egyptians could not drink the water from the river".*

The first recorded human poisoning due to toxic phytoplankton was observed in 1793 in the area of the British Columbia, now known as Poison Cove. The algal toxins are alkaloids that cause Paralytic Shellfish Poisoning (PSP) even at doses as low as 500 µg. Every year, the authorities around the world report 2,000 cases of human poisoning from the consumption of fish or shellfish while 1.5% of these cases are lethal. In addition to humans, whales and porpoises have been victims of toxic phytoplankton through consumption of fish or zooplankton (Geraci et al., 1989). There are two main categories of harmful algal blooms: (a) species producing harmless water discoloration. These species under exceptional conditions of density can cause deaths to fishes and invertebrates through oxygen depletion and (b) toxins can be either neurotoxins or gastrointestinal toxins. The types of different toxins, toxic species and relevant references are given in Table 1.1.

Although algal bloom phenomena have been known since ancient times, it is during the last few decades that the frequency of algal bloom events has increased to an alarming level. Most of these events have a serious economic impact on both biological resources and recreational activities. Paralytic fish poisoning shows a geographic wide distribution along the East and West coasts of U.S.A., in coastal areas of the Northern European countries, in the Eastern Mediterranean, many Southeast Asian areas and a few areas in Australia (Hallegraeff et al., 2003). Species of the genus *Alexandrium* are widely spread, almost cosmopolitan in Europe, North America, Japan but also in many places in the Southern Hemisphere.

Although the causes for algal bloom formations have not been documented well so far, there are a number of factors that seem to favor and/or trigger toxic bloom events: (a) eutrophic conditions in coastal waters due to cultural eutrophication sources (b) scientific knowledge and awareness about toxic species (c) climate change (d) transportation of toxic species through the ballast water and (e) keeping better records on algal toxic events due to economic and social impact as well as the attention of social media and press. Literature on toxic algal blooms increased exponentially since the 70s. These days more than 500 papers are published per year on the subject.

According to pathologists there are four classes of toxins produced by microalgae (Segar, 1997): toxins that cause Paralytic Shellfish Poisoning (PSP), Neurotoxic Shellfish Poisoning (NSP), Diarrheic Shellfish Poisoning (DSP) and Amnestic Shellfish Poisoning (ASP). Two *Ichthyotoxic microalgal* species *Pfiesteria piscicida* and *Pfiesteria shumwaye* were identified during 1991 in North Carolina (Burkhoolder et al., 1992). The fish killing was characterized mysterious at the beginning because of their ephemeral physiological activity. These algae appear in the form of cysts that germinate once it happens to be within live fishes and they encyst again after the fish death. *Pfiesteria* is now spread worldwide in New Zealand, Australia and Europe.

In spite of the fact that bloom forming mechanisms are not known as yet, there are cases however, that cultural eutrophication seems to favor algal blooms as a

Table 1.1 Types of effects from toxic microalgal blooms.

Type of effect	Examples of microalgal toxic species	References
Paralytic Shellfish Poisoning (PSP)	*Alexandrium catenella* *Gyrodinium catenatum* *Pyrodinium bahamense* var. *compressum* *Alexandrium cohorticula*	Morse et al. (2018) Hallegraef and Fraga (1997) Deo Florence et al. (2014) Ogata et al. (1990)
Amnestic Shellfish Poisoning (ASP)	*Pseudo-nitzschia australis* *Pseudo-nitzschia delicatissima* *Pseudo-nitzschia multiseries* *Pseudo-nitzschia pungens* *Pseudo-nitzschia seriata*	Schnetzer et al. (2017) Lelong et al. (2014) Lewis et al. (2018) Tenorio et al. (2016) Fehling et al. (2004)
Neurotoxic Shellfish Poisoning (NSP)	*Karenia brevis* *Karenia papilionacea* *Karenia selliformis* *Karenia bicuneiformis*	Dorantes-Aranda et al. (2015) Yamaguchi et al. (2016) Feki et al. (2013) Botes et al. (2003)
Diarrheic Shellfish Poisoning (DSP)	*Dinophysis acuta* *Dinophysis acuminata* *Dinophysis caudata* *Dinophysis fortii* *Dinophysis norvegica* *Dinophysis rotundata* *Dinophysis sacculus* *Prosocentrum lima*	Swan et al. (2018) Paterson et al. (2017) Basti et al. (2015) Uchida et al. (2018) Blanco et al. (2015) Morton et al. (2018) Bazzoni et al. (2018) Moreira-Gonzalez et al. (2019)

number of species have been identified as nutrient prone. An exponential increase in algal blooms in Hong Kong during 1976–1986 (Lam and Ho, 1989) dominated by *Karenia mikinotoi, Gonyaulax polygramma, Noctiluca scintillans* and *Prorocentrum minimus* was well correlated to a six-fold increase of human population during that period and 2.5 fold increase in the nutrient loading. Similar trends were also observed in Japan (Seto Island Sea—a major fish farming area) during the period 1965–1976 (Okaiki, 1989): red tide outbreaks of *Chattonella antiqua* and *Karenia mikinotoi* showed an increase in COD from sewage and pulp paper factories effluents. Algal bloom episodes related to nutrient increases have also been observed in the North Sea in Europe (Smayda, 1990) as well as in the Dutch coastal waters (Lancelot et al., 1987). Cultural eutrophication, even if it not the triggering mechanism of toxic algal blooms, seems beyond any doubt to support their formation. It is therefore obvious that the eutrophication impact in this case is both economic and social, especially in countries depending a great deal on mariculture. Furthermore, "tailored" monitoring programs should be designed and implemented, focusing on frequent sampling for harmful algal species detection and enumeration in the water column as well as analytical work on their toxins.

1.9 Atmospheric Deposition of Nutrients

Until the 80s, there was a general belief that the main sources of nutrients in the marine environment were land-based sources (river discharges and non-point nutrient runoff), resuspension of sediments and nutrient recycling (mainly nitrogen and phosphorus cycles). It is only from the 90s that atmospheric deposition was characterized as a significant nutrient source, especially in the oceans where distance from terrestrial areas and depths, substantially restrict the land-based and resuspension mechanisms (Duce et al., 1991; Krishanamurtry et al., 2007; Krom et al., 2004, 2010; Okin et al., 2011).

The significance of the atmospheric deposition was indicated by Lazzari et al. (2011) who used a 3-D marine ecosystem model simulating plankton chlorophyll and primary production in the pelagic Mediterranean Sea; the results showed that atmospheric and terrestrial nitrogen sources account for less than 5 per cent of the total integrated primary production on an annual basis. Inputs of anthropogenic nitrogen in the sea through atmospheric deposition processes, have shown considerable increase since 1960 because of intensified agricultural practices and fossil fuel substances (Christodoulaki et al., 2013). Nitrogen deposition in 1960 was 50 $mgNm^{-2}y^{-1}$ with only a few exceptions exceeding 200 $mgNm^{-2}y^{-1}$; this deposition mainly came from natural sources. At the beginning of the 21st century, atmospheric deposition of nitrogen in the oceans exceeded 200 $mgNm^{-2}y^{-1}$ and in some cases reached 700 $mgNm^{-2}y^{-1}$ (Duce et al., 2009). The anthropogenic influence on reactive nitrogen deposition is so dramatic these days that it is approaching the levels of molecular nitrogen fixation. One view (Mahoward et al., 2008) is that increased rates of nitrogen deposition of anthropogenic origin in the ocean can change the nutrient regime from nitrogen-limited to phosphorus-limited marine environments. Deposition plumes spread downwind of major urban cities off the Asian, Indian, American (South and North), African and European continents (Duce et al., 2008).

Although it is now clear that atmospheric inputs significantly influence ocean productivity (Jickells et al., 2018), the mechanisms are not sufficiently understood. However, it has been suggested that anthropogenic nitrogen inputs can lead to phosphorus limitation (Fanning, 1989). Furthermore, nutrient availability in the ocean enhances productivity which in turn influences the carbon dioxide storage in the marine ecosystems. Deposition of sufficient amounts of iron may stimulate growth of diazotrophic microorganisms, leading to atmospheric nitrogen fixation and nitrogen availability through atmospheric deposition and stimulates non-diazotrophic organisms also enhancing primary productivity. Both mechanisms can finally lead to phosphorus limitations. It is obvious that there is a strong interconnection among nitrate, phosphate and iron nutrients in the marine environment. Atmospheric deposition of nitrogen changes the nutrient balance in the marine environment, influencing primary productivity and phytoplankton ecology.

The second nutrient contributing to primary productivity after nitrogen, is phosphorus. There are estimates that the net gain of the total phosphorus of the oceans from terrestrial sources is about 560 GgPy^{-1} (Mahoward et al., 2008). The main mechanisms for atmospheric deposition of phosphorus worldwide are mineral aerosols (82 per cent), followed by primary biogenic particles (12 per cent) and combustion sources (only 5 per cent). The last mechanism is important for non-dusty regions. Phosphorus combustion through atmospheric processes can be vital to oligotrophic environments where primary productivity is often phosphorus limited. Iron is a trace element for the growth of phytoplankton, occurring at concentrations of the order of μg/l but it is absolutely necessary for algal physiology (Fogg and Thake, 1987) and can be a limiting nutrient in primary productivity. For the open ocean, atmospheric deposition is the primary source of iron; this is mainly due to the fact that iron of riverine origin tends to be deposited in coastal sediments (Jickells et al., 2005).

In spite of the huge area covered by the oceans (about 1.5 billion km^2), sixty per cent of that area is depleted in nitrate and phosphate sustaining phytoplankton growth and primary productivity at low levels. These "ocean deserts" are primarily the central ocean gyres. The oceanic oligotrophic regions, also referred to as Low Nutrient Low Chlorophyll (LNLC) areas, are characterized by chlorophyll concentrations less the 0.07 mgm^{-3} and they are dominated by small phytoplankton and heterotrophic bacteria (Cho and Azam, 1990; Uitz et al., 2010; Guieu et al., 2014). Atmospheric deposition of nitrogen studies in the marginal East Asian Seas (Taketani et al., 2018), indicated that the levels of ammonium-nitrogen atmospheric inputs were almost the same as those from river outflows. The same authors have estimated that biological primary productivity is increased by 1.1–3.9 per cent due to atmospheric nitrogen inputs. It is therefore obvious that atmospheric deposition cannot be neglected as a source of nutrients.

Atmospheric deposition also occurs in regional seas. A typical example is the Mediterranean Sea which is extremely poor in nutrients regarding the surface waters: frequent injections of nutrients from the Sahara desert, enrich the Mediterranean Basin. Nutrient deposition may explain 1 to 10 per cent of the seasonal chlorophyll variations in the low nutrient areas of the Mediterranean (Gallisai et al., 2014).

Nutrient contributions from deserts may become larger due to an increasing desertification trend and the shallowness of the mixed layer.

Atmospheric nutrient deposition has not been connected so far with eutrophication problems. However, the underlying mechanisms should be taken into account in phytoplankton studies, including eutrophication as they have become the primary regulating factor, especially in offshore waters.

1.10 Ecosystem Services and Eutrophication

By the term Ecosystem Services, we mean ecosystems' and species' functions and processes that support sustainability and fulfill human life (Garpe, 2008). Although humans were using these services at the dawn of the mankind, it is relatively recently, the late 1990s that ecosystem services enjoy the attention they deserve by the scientific community. This is because it was then that monetary values related to ecosystem services were actually conservatively estimated. This step highlights the economic and societal values of ecosystem services that were, until that time, unvalued and usable by anybody in an unsustainable way, sometimes against other natural resources and services (United Nations, 2017). The marine environment is not only a large reservoir of ecosystem services, not quantified in most cases, but also vital to humans, especially if we take into account that 44 per cent of the world's population lives on or within 150 kilometers from the coast (van der Belt et al., 2017). Ecosystem services have been classified into four categories: provisioning services, regulating services, cultural services and supporting services (Everard, 2017; Millenium Ecosystem Assessment, 2005). Provisioning services include food suitable for consumption, chemical resources, genetic resources, energy, space and waterways. Regulating services are sediment retention, mitigation of eutrophication, regulation of toxic substances and climate regulation. Cultural services are variable, also depending on human attitudes but they are mainly recreation, cultural heritage, inspiration (such as painting and music), scenery, science and education as well as legacy on the sea. By supporting services, we mean biogeochemical cycles, primary productivity, food webs, ecosystem's resilience, habitats and diversity.

The 23 ecosystem services mentioned above are the most common in the marine environment. They form a complex system, being the result of climate, chemical, physical, biological and geological processes. Sustainability can be thought of, if anthropogenic influence ranges within system's resilience. Marine eutrophication of human origin is a powerful environmental component that can affect many kinds of ecosystem services and their interactions. It is not within the scope of this book to analyze the relationships between marine eutrophication and ecosystem services. Nevertheless, many impacts of eutrophication mentioned in previous sections, are connected with ecosystem services (oxygen depletion, diversity, toxic algal blooms, etc.). The focus of this section will be to draw reader's attention to the interlinks between effects of eutrophication and ecosystem services.

Marine ecosystem services are vulnerable to eutrophication because any changes in ecosystem's community structure and species, plays a key role in regulating processes that can have a serious impact on both the marine environment and the society (Luck et al., 2009). Although the simplest way for quantification of ecosystem services is to measure their economic values, this is not always possible, especially when eutrophication-induced changes are considered (Alexander et al., 2016). This is particularly difficult when many ecological changes occur at the same time and when in addition to services connected with income (for example, fish catchments) there are also non-material benefits and culture services: these include the availability of some local produce or the preservation of fishing culture, both difficult to quantify.

Seagrasses provide many supportive ecosystem services including sediment stabilization, erosion prevention, water flow reduction, nutrient and organic material trapping and habitats suitable for fish and invertebrates. They are also a good substrate for epiphytes and periphyton, both sources of food for herbivores. Due to the ecological services they provide to coastal systems, they are globally considered among the most valuable ecosystems (Adams, 2016; Costanza et al., 2014; Short, 2007). However, excessive nutrient supply leads to an overgrowth of epiphyte and nuisance to seaweed populations; these factors exercise pressure on the seagrass assemblages, competing with submerged macrophytes for light and therefore, reduce both their coverage and biodiversity. There are two alternatives regarding eutrophication impacts on seagrasses: (a) nutrient rich situation, may favor excessive growth of macroalgae and (b) high phytoplankton density may exclude in the end rooted of submerged macrophytes (Flindt et al., 1999; Viaroli et al., 2008).

Macroalgal systems are complex ecosystems providing services such as shelter, protection and food for many animal species. However, excessive growth of macroalgae due to excessive nutrient supply can lead to algal bloom formations that can cause hypoxia, release hydrogen sulphide into the sediment and loss of species with ecological and economic importance (Lyons et al., 2014). Overfishing changes top-down control and can drive down, in some cases, populations of invertebrate grazers, leaving room to microalgae to proliferate. This synergistic interaction between herbivore activity and nutrient control through primary producers, has already been dealt with (Burlepile and Hay, 2006). Opportunistic macroalgae are favored in a nutrient rich environment to form dense canopies, mats of algae or settle on the bottom. Irrespective of the type of growth, macroalgal blooms have negative impacts on a number of ecosystem services: they inhibit recreation, downgrade scenery, impact tourism, relegate fisheries and can also be a problem to mariculture (de Leo et al., 2002). Intense macroalgal proliferations may alter community structure and processes of the ecosystems in the area (Raffaelli et al., 1998). Toxic gases released from macroalgal assemblages, mainly rotting plants can also be a health hazard to humans (Lyons et al., 2014).

On the other hand, macroalgal blooms can have some positive effects regarding ecosystem function. Increased habitat complexity facilitates dispersal of other species and provides shelter and food to animals (Wilson et al., 1990). From this point of view, macroalgal overgrowth increases biodiversity and secondary production in these ecosystems (Bolan and Fernandes, 2002; Dolbeth et al., 2003). Top down and bottom up control processes in seagrass and seaweed communities have been recently studied by Ostman et al. (2016). The interlink between eutrophication and ecosystem services has been apprehended by O'Higgins and Gilbert (2014); using the North Sea as a case study area, they related ecosystem services and eutrophication to Good

Environment Status, required by the Marine Strategy Directive (MSFD). Ecosystem services are now an essential element in the Ecosystem Approach management. In addition to the 11 Environmental Descriptors of the MSFD, this approach, although promising, does not seem to be always feasible: the authors are of the opinion that in many North Sea areas the data is not sufficient to support the Ecosystem Approach by valuating ecosystem services.

Rhodes et al. (2017) have tried to quantify the connection between eutrophication and ecosystem services. They have identified 13 Ecological Production Functions within different ecosystems, including estuaries, and connected changes in phytoplankton to ecological end-points, offering benefits that both the public and policy makers can appreciate. The results have stressed the importance of the interlinks between the natural and social sciences as well as the need for joint efforts when ecological stressors are assessed.

1.11 Climate Change and Eutrophication

Even natural eutrophication, that is eutrophication without human or cultural influences in the marine environment, is a complex process, extremely variable, depending on site factors such as nutrient stoichiometry, geomorphology, biodiversity and climate change. As our planet undergoes many changes, stemming from the new climate regime, acceleration of eutrophication, catalyzed by new global climate conditions, becomes a matter of concern not only for ecologists but also for engineers, water quality scientists, managers and policy makers (Ansari et al., 2011). There is already apprehension among scientists and policy makers involved in the setting up, implementation and effectiveness of the evaluation of measures on eutrophication in regional conventions for the protection of the marine environment, that the measures proposed may be inadequate if climate change processes are involved. Doubts have already been expressed that strategic objectives set for the Baltic Sea by the Baltic Sea Action Plan (BSAP) for "*a Baltic Sea unaffected by eutrophication*" will finally be unattainable under a climate change. In spite of the stringent measures adopted, eutrophication will continue to be the most crucial problem in the Baltic (WWF, 2008). It is, therefore, a new challenge to reset eutrophication policy worldwide in order to respond to the threat of enhanced eutrophication during the process of global climate warming. Any new or revised policy on this subject, assumes understanding of the processes linking eutrophication to climate warming; a short account of the processes, concerns and specific aspects of marine eutrophication under warming conditions will be given in the present chapter.

The principal characteristic of climate change is increased air temperature. Earth's temperature has increased during the 20th century approximately by 1°C (IPCC, 2007), whereas during the current century, global temperatures are expected to go up by an additional 1.5–5.0°C (IPCC, 2007). Due to sea-atmospheric interaction, the sea surface temperature is bound to increase, promoting photosynthesis. However, there are plants that cannot cope with high light intensities mainly to the ultraviolet part of the spectrum, setting therefore selection criteria on marine plant species (Hughes et al., 2012). It is well known that natural communities of phytoplankton are influenced by higher temperatures as algal growth rates are temperature dependent (Goldman and Carpenter, 1974; Raven and Geider, 1988). A shift that has been observed is the succession of bacillariophyceae, mainly diatoms (< 20°C) to flagellates (20–25°C) and finally to cyanophyceae, Hunter (1998), a phylum of blue green algae. This way diversity is reduced and at the same time metabolic rates and oxygen demand increase.

Nutrient availability in connection with increased surface temperatures, favor the proliferation of cyanobacterial blooms (Peperzak, 2003). Increased temperatures not only accelerate metabolic processes but also affect various other processes that contribute to algal growth as they change many physical characteristics of the aquatic phase. Higher temperatures decrease viscosity and therefore facilitate nutrient diffusion towards the cell surface. This is a particularly important process as species succession depends, to a great extent, on nutrients (Dugdale, 1967). As the buoyancy of the unicellular organisms in the marine environment is dependent on the viscosity, higher temperatures decrease the viscosity, promoting sinking of larger cells; this way offering an advantage to small size microalgae to dominate the surface layer (Walsby et al., 1977; Wagner and Adrian, 2009). Another indirect effect due to increased surface temperature is the strength of stratification. When the surface layer is isolated from deeper water masses, it soon becomes nutrient deprived, leading to nutrient limitation. Nutrient limited conditions favor cyanobacteria because they can either fix dissolved nitrogen (diazotrophic cyanobacteria) or move to deeper water masses by regulating their buoyancy (Paerl et al., 2008). Climate change may also affect salinity values in estuarine systems: the rise of the sea level, reduced river outflow due to a prolonged period of drought or even increased river outflow due to stormy conditions, all these induce fluctuations in salinity. This can lead to a restructuring of the estuarine phytoplankton communities (Moisander et al., 2002; Bordalo and Vieira, 2005).

As dinoflagellates have mainly been studied so far for algal bloom formation (Hallegraef et al., 2003), not much attention has been paid on marine algal blooms dominated by cyanobacteria. The most common genera of marine cyanobacteria involved in marine algal blooms are *Lyngbya*, *Trichodesmium* and the coccoid cyanobacterium *Synechcoccus* (O'Neil et al., 2012). Species of *Lyngbya*, although they are benthic forms growing on seagrass, macroalgae and corals, they may form intense algal blooms near the surface once they are detached with serious economic and social impact. Blooms of *Lyngbya majuscule* are these days abundant in many tropical and subtropical regions; these blooms have been connected to anthropogenic eutrophication. Bloom events of *L. majuscule* in a place of the Australian Great Barrier Reef were recorded when an observation platform was installed and the environment was polluted due to nutrient from the droppings of many birds roosting the platform (Albert et al., 2005). *Trichodesmium* is a colonial species thriving in tropical and subtropical regions where blooms are formed fairly often and can thrive for weeks causing problems, not only because of their high density and potential toxicity but also during the bloom decay. Although *Trichodesmium* is well distributed in low nutrient environments, nutrient enrichments due to anthropogenic influences such as land clearing or even Fe-rich Aeolian dust deposition, connected with desertification, seems to favor its growth (Lenes et al., 2001). *Trichodesmium* seems to spread geographically, following climate change in the marine environment. *Trichodesmium erythreum*, a characteristic species of the Red Sea was identified in the Aegean waters (Mediterranean) for the first time, a few years ago (Spatharis et al., 2012). *Synechococcus* is a cosmopolitan open ocean species, also forming algal blooms in multiple ecosystems. These blooms may cover areas as large as 100 km^2 and last for months. Elevated nutrient concentrations have been identified as the primary causes of *Synechococcus* blooms in Florida Bay; however, this aspect is now disputed as high nutrient concentrations also favor the growth of larger phytoplankton (Rave and Kubler, 2002).

The facts presented above indicate that there is a strong interconnection between climate change and eutrophication, not only because higher temperatures as well as higher nutrient concentrations induce high growth rates but also because of many processes that are driven indirectly by climate change favoring phytoplankton growth in general and frequent bloom formations.

2

Data Analysis Methods for Assessing Eutrophication

2.1 The Need for Data Analysis

The difficulty in quantifying eutrophication is due to the complexity of the ecosystemic processes along with the diversified pollution sources inducing marine eutrophication, impacting the marine environment, especially the human resources which include fish food and recreation. The severity of the problem has stimulated international organizations and coastal states to support monitoring and research on this problem. Intensive research worldwide has led to the accumulation of an enormous amount of data. National and international marine monitoring programs (Karydis and Kitsiou, 2013), monitoring research projects of local interest, surveys following red tide or even toxic algal bloom episodes, mesocosm (Petersen et al., 2009) and microcosm experiments (Karydis, 2014) usually designed to study ecological mechanisms influenced by nutrient enrichments as well as enrichment experiments using algal cultures have significantly contributed to the understanding, quantification and assessment of the eutrophication phenomenon. However, all this vast amount of data is poorly understood unless data processing is carried out to respond to specific questions. In addition, policy makers require simple and short answers so as to make decisions regarding measures that need to be taken. Policy makers find it difficult to incorporate the outcome of research findings in the legislation due to uncertainties, conflicts and systemic complexities. On the other hand, marine scientists are reluctant to compress and curtail the information to a "good" or "bad" outcome, even if they can use a scale, usually an ordinal scale, such as bad, good, very good grades common in recent European Union Directives (EC, 2000) or other legislative measures.

In addition, there are drawbacks concerning the methods used for data analysis. There is also a dispute as far as the ecological reliability is concerned to "provide a basis for practical action in environmental problems" (de Jonge, 2006). These problems require data analysis approaches that include multivariate analysis methods of ecological communities (Digby and Kempton, 1987; Krebs, 1999), ecological indicators (Karydis, 2009; EC, 2000; Borja and Dauer, 2008). Geographical Information Systems (GIS) applications (Kitsiou et al., 2002). Mutliple Criteria Analysis (MCA) methods (Janssen, 1992; Kitsiou and Karydis, 2000) and simulation modelling (NRC, 2000).

Especially in the European regional seas, an effort has been made over the last thirty years to assess eutrophication levels. This effort has resulted into the completion of three targets that form the background for developing the legal framework aiming at the implementation of management measures to reduce nutrient loading and ensure good coastal water quality. These targets include: (a) Systematic data collection, mainly through long term monitoring projects; these data sets allow the assessment of the trophic status in the European seas and detection of possible trends (b) the use of statistical methods and indicators that can provide reliable information on the trophic status and (c) the development of numerical models linking nutrient loading with physical and biochemical cycles (Ferreira et al., 2011).

2.2 Data Collection

Due to the large number of direct and indirect effects, a number of variables are measured when eutrophication problems are encountered. The variables directly involved in marine eutrophication can be clustered into two groups: the causes and the effects. As causes, inorganic nutrient concentrations, namely orthophosphate, nitrate, nitrite, ammonium and silicates are measured. The main variables expressing direct effects are chlorophyll concentrations, phytoplankton cell number, species composition and their relative abundance in the samples. As background information, temperature, pH, water transparency and salinity are required. The monitoring strategy set up by UNEP within the MED POL framework (UNEP, 2003a) recommends that in addition to the variables mentioned above total phosphorus, total nitrogen and dissolved oxygen should be considered as mandatory variables. Measurements on the prevailing currents and water mass dynamics can provide an estimate of the residence time especially in semi-enclosed areas and lagoons. Over the last two decades, benthos was also considered as an important element in understanding and assessing marine eutrophication. Species composition of phytobenthos, meiobenthos and macrozoobenthos can provide information on population dynamics and levels of eutrophication especially in shallow coastal waters (< 30 m).

Over the last decades there have also been two supplementary techniques that provide enormous amounts of data in eutrophication studies have been also available, that is, remote sensing and automatic buoys. Remote sensing methods can be useful to study spatial distributions of variables connected with eutrophication. They can be applied on chlorophyll, turbidity and suspended matter distributions. Maps can be generated by integrating remote sensing data with *in situ* measurements. Remote sensing can also be used for monitoring either to create a series of historical images of an area enabling the researchers to detect the development of eutrophic trends or to optimize the *in situ* network of sampling sites (Kitsiou et al., 2009). Remote sensing at a sub-regional or even local scale can also provide spatial information on algal blooms and their fate (UNEP, 2003a).

The use of automatic buoys increases and they are particularly useful in eutrophication studies. They provide information on a time series basis of dissolved oxygen and chlorophyll concentrations (via fluorescence measurements). Lately, it is possible to measure some of the nutrients as well. In spite of problems related to fouling of equipment, especially the sensors, automatic buoys either on fixed or drifting platforms seem to gain ground over the standard sampling procedures. A sophisticated system based on automatic buoys for monitoring algal blooms is the MARINET system (Hallegraeff, 2003). This system has been developed in Norway and deplored along the Norwegian coasts and fiords where aquaculture units exist. It is actually an early warning system for algal bloom detection allowing the aquaculturist to tow the cages away. The importance of such a data collection system becomes obvious if we take into account the fact that in 1988 during the *Chlorochomulina* blooms, 1800 cages containing about 26,000 tons of fish were removed into safe areas.

Eutrophication studies these days form part of integrated approaches aiming at marine policy (de Jong, 2006; Karydis and Kitsiou, 2012; Karydis, 2015), coastal management (Turner and Bateman, 2001) or even Marine Spatial Planning (Kitsiou et al., 2017). It is now a common practice that in most cases information concerning socio-economic aspects is collected at the same time. The catchment area (area, topography, natural drainage and possible land use modifications) are recorded. In addition, information on waste water discharges and abstraction, port and harbor areas, shipping routes, anchorages, pipelines and cables, fishing activities, aquaculture facilities, marine wind parks and tourism (number of visitors/tourist facilities) is collected (Shucksmith et al., 2004).

2.3 Shortcomings in Data Analysis

The analysis of inorganic nutrients, chlorophyll α concentrations and phytoplankton biomass are based on standard routine procedures (Parsons et al., 1989; Vollenweider, 1974) and they can be carried out by an average level laboratory, at low cost and without any special expertise. However, nutrient variables that is nitrate, nitrite, ammonia, phosphate and silicate, chlorophylls, phytoplankton, cell numbers and ecological indices derived from the qualitative/quantitative analysis of phytoplankton communities are intercorrelated (Ignatiades et al., 1985; Primpas et al., 2010a; Kitsiou and Karydis, 2011). In practical terms this means that the total effect is not additive; however, some of the variables are uncorrelated and it is confusing that there is negative correlation between some of the variables (Ignatiades et al., 1985).

Another problem regarding the "behavior" of the variables is their frequency distributions. It is well established by now that the nutrient variables usually follow the log-normal distribution (Ignatiades et al., 1992) whereas chlα values seemed to follow the gamma distribution. This is also supported by Heyman et al. (1984) who analyzed 950 samples from 25 meso and hypertrophic lakes in Sweden. Ignatiades et al. (1992) studied the frequency distribution of phosphate, nitrate and ammonia samples from Saronicos Gulf, Greece. Sample collection was carried out from four sub-systems of the Gulf over a 24 months period: 347 samples from inshore waters, 595 from offshore waters and 120 samples from nearshore pelagic water and 197 samples from offshore pelagic waters. In all cases the variables followed the log-normal distribution.

The annual cycle of nutrient concentrations and phytoplankton cell number is characterized by seasonal trends (Pagou and Ignatiades, 1988; Ignatiades et al., 1986). That means that high variations around the mean annual values as well as overlapping among variable distributions representing oligotrophic, mesotrophic and eutrophic conditions complicate the assessment (Giovanardi and Tromellini, 1992; Ignatiades et al., 1992). As nutrient and phytoplankton are built-in variables in the marine ecosystem, it is not always easy to discriminate between the natural and anthropogenic sources of nutrients as well as the proportion of phytoplankton biomass induced by human activities. The concentrations of nutrients causing eutrophication depends on many factors beyond nutrient inputs: the topography of the area, the depth (if we refer to coastal systems), the resuspension of nutrients from the sediment, vertical water movements, especially upwelling and transportation due to currents are the most predominant factors that need be taken into account.

Selection of reference sites is necessary when built-in variables are involved so that any measurements in the survey area can be compared to the concentrations from the reference site, that is the background values. It is not always easy to select reference sites: if they are near the survey area they may be influenced by the nutrient loads, if they are located further away they may characterize a different ecosystem. In addition, reference values differ from one region to another: a good example is the comparison of reference values between the oligotrophic Mediterranean and the eutrophic areas of the North or the Baltic Sea (EEA, 1999).

2.4 Indicators for Assessing Eutrophication

The initial use of "indicators" in the marine environment was to describe different aspects of the marine ecosystem. Most of them were condensing information regarding species abundance, species dominance and species diversity. Later, it was realized that some indices could also be used for setting water quality criteria (Whihm and Dorris, 1968). Washington (1984) has given a comprehensive review on diversity, biotic and similarity indices with special relevance to ecosystem quality. Ferris and Humphrey (1999) have reviewed the potential of bioindicators in environmental quality studies. Meanwhile a number of indicators have been developed, beyond the ecological indices that have been applied in fields such as marine pollution, coastal management and lately in marine spatial planning (Kitsiou and Karydis, 2017). According to Walz (2000) "*an indicator is a variable that describes the state of the system*". This definition may be the simplest definition of indicator found in literature. A definition on "indicator" relevant to ecological and water quality status has been proposed by Humphrey (1999): "*An indicator may be defined as a characteristic which, when measured repeatedly, demonstrates ecological trends and a measure of current state of quality of an area*". A broader definition on indicators has been proposed by Burger (2006): "*indicator: index or measurement end point to evaluate health of a system: physical, biological, economic and human"*. However, some popular indicators such as the Shannon's species diversity index should also be used with caution (Spatharis et al., 2011) in assessing levels of eutrophication as there are a number of shortcomings: (a) as mentioned above defining reference conditions is not always possible (Nielsen et al., 2007; Borja et al., 2006). (b) richness and diversity indices have been developed to describe ecological aspects of an ecosystem; their use as environmental indicators can be erroneous (Karydis and Tsirtsis, 1996). Invasive species although contribute to systems' diversity, do not necessarily improve the environmental quality. Sometimes the "invaders" in phytoplankton communities are toxic microalgae: in this case the deficit of an indicator to express environmental quality is more than obvious. In addition, indices should consistently increase or decrease along with environmental gradients, in other words they should show monotonicity, a feature that does not always occur in their use as water quality indicators. The monotonicity of 22 ecological indices was investigated using simulated data in an effort to depict some indicators suitable for developing an ecological quality scale using phytoplankton data. It was found out that only three out of the 22 ecological indices tested (Tsirtsis and Karydis, 1998), namely Margalef's, Menhinick's and Evenness E_1 indices, increased monotonicity. Margalef's index expresses species richness and is based on a linear relation between species number and the logarithm of the number of phytoplankton cells in the sample. Menhinick's index is also expressing species richness but shows a smaller sample variation. The Evenness E_1 index is a measure of equality in species abundance in the sample. Detailed information on those indices can be found in Magurran (2004), Washington (1984), Karydis and Tsirtsis (1996) and Karydis (2009).

The effectiveness of ecological indices in assessing eutrophication has been investigated using multivariate procedures (Karydis and Tsirtsis, 1996; Danilov and Ekelund, 2001). Only three out of twelve ecological indices were found suitable for assessing eutrophic levels (Karydis and Tsirtsis, 1996): species number, total number of individuals and Kothe's species deficit. This species deficit index has been proposed as an estimate of species deficit between a discharge area and a reference site (control site). On the other hand, Shannon's index, the most popular index in literature did not respond to eutrophic levels (Karydis and Tsirtsis, 1996). A number of indices that is Shannon's H' and D', Simpson's, Hill's N_1 and N_2 showed a characteristic hump-shape curve (Spatharis and Tsirtsis, 2010). An upward trend was observed at low abundance values which was less than 10^4 cells/l, followed gradually by a decrease reaching their minimum at very high abundance values. It was found that only the Evenness Indices E_2, E_3, E_6 and the Menhinick's Index fulfilled the two criteria of monotonicity and linearity.

A eutrophication index based on nutrient concentrations has been proposed by Karydis et al. (1983). It is based on nitrate, nitrite and ammonia concentrations in an area. The index requires calculation of the total annual concentrations of the nutrients in the area and the total annual nutrient concentrations of the station under assessment. The index is given by the formula:

$$I = \frac{C}{C - logx} + log A \qquad (1)$$

Where I is the index of eutrophication, C is the log of the total annual concentration of the nitrogen nutrients and x is the total nitrogen nutrient concentrations of the station under investigation.

A multimetric index that becomes rather popular in eutrophication studies with the time is the TRIX index proposed by Vollenweider et al. (1998) and adjusted for oligotrophic environments by Primpas et al. (2011). This index is a linear combination of four physical, chemical and biological variables namely chlα, dissolved organic nitrogen, total phosphorus and the absolute percentage of deviation of oxygen from oxygen saturation values. The scale of eutrophication proposed, based on samples collected in the Adriatic Sea was incorporated in the National legislation of Italy as an index officially used for assessing environmental quality (Parlamento Italiano, 1999). The TRIX index has also been used for assessing eutrophication in the Caspian Sea (Nastollahzadeh et al., 2008; Shahrban and Etemad-Shahidi, 2010), the Persian Gulf (Taebi et al., 2005), the Black Sea (Moncheva et al., 2001), the Montego Estuary (Salas et al., 2008), the Southeast Mexico (Herrera-Silveira and Morales-Ojeda, 2009) and the Helsinki Sea (Vascetta et al., 2008). In spite of the extensive use of TRIX index, there are drawbacks regarding the suitability of the index: (a) it has been reported (Kitsiou and Karydis, 2011) that *"the TRIX index does not conform with the requirements of the Water Framework Directive WFD 2000/60/EC (2000) as the scale is not normalized to local reference conditions and physical, chemical and biological variables are built-in within the index, although the Directive requires interaction of separate evaluations of the above mentioned variables"* (Pettine et al., 2007) and (b) the bottom of shallow coastal waters is dominated by both macroalgae and some phanerogam species that contribute a great deal to the nutrient dynamics in those transitional waters (Giordani et al., 2009). As the reference conditions in most of the regional seas mentioned above are characterized as eutrophic, they differ from the reference values in oligotrophic environments such as the Eastern Mediterranean. Primpas et al. (2010) have attempted to rescale the TRIX index. This approach was followed by researchers in various marine areas (Pearman et al., 2018; Simboura et al., 2016; Kong et al., 2017).

In addition to the indices presented above, a composite Integrated Phytoplankton Index (IPI) was proposed based on phytoplankton metrics which is the sum of abundance values, diversity values and chlα concentrations (Spatharis and Tsirtsis, 2010). The weak point with this index is the fact that the authors have not explained why the three variables used should have the same weight. A multimetric index efficient in discriminating oligotrophic, mesotrophic and eutrophic waters has been proposed by Primpas et al. (2010). This index named "Eutrophication Index" (E.I.) is based on four nutrient variables and chlα concentrations. The index was tested using three standard sets of data characterizing eutrophic, mesotrophic and oligotrophic water types (Ignatiades et al., 1992). The proposed formula for E.I. is:

$$E.I. = 0297C_{PO4} + 0.296C_{NO3} + 0.275C_{NH3} + 0.214C_{Chl\alpha} \tag{2}$$

where C: chlα and nutrient concentrations.

Principal Component Analysis was used for the derivation of the coefficients. The E.I. was validated using sets of data from the area of Rhodes, Greece (Karydis,

1996) and it was found that the results were compatible with previous work (Karydis, 1992; Moriki and Karydis, 1994; Karydis, 1996). Index scaling was adapted to five classes to conform to the requirement of the European Water Framework Directive: High quality: less than 0.04, Good: 0.04–0.38, Moderate: 0.38–0.85, Poor: 0.85–1.51 and Bad: greater than 1.51.

It has already been mentioned that extensive seagrass formations in the shallow systems use considerable amounts of nutrients from both the water column and the sediment (Zardivar et al., 2008). Benthic communities are also affected by phytoplankton and periphyton dynamics in both the water column and the sediment. Changes induced to the invertebrate benthic community provide valuable information on water quality as some of these organisms have a long life cycle which responds to anthropogenic stress (Pearson and Rosenberg, 1978). Their role in assessing environmental quality is obvious (Elliot, 1994) and therefore a number of indicators have been proposed for assessing trophic states. The Ecological Evaluation Index (EEI) based on macroalgal and phanerogam taxa, expresses the ecological quality of coastal and transitional waters. It ranges from 2 to 10 and classifies the marine ecosystems from high to bad (Orfanidis et al., 2001, 2003). The AMBI Index (biotic index) based on data from macrobenthos, has been proposed by Borja et al. (2000) to characterize the ecosystem quality of European coastal waters. A combination of the AMBI index with Shannon's diversity index is known as M-AMBI. Simboura and Zenetos (2002) have proposed the BENTIX Index. BENTIX is using data from the macrozoobenthos soft substrate and it is suitable for ecological quality assessment conforming to the requirements of WFD. The species are clustered into three ecological groups depending on their response to organic pollution. The ISD index based on a taxonomic free method uses the way individuals from benthic communities are distributed over biomass size classes (Reizopoulou and Nikolaidou, 2004). The skewness of the distribution is an estimate of community disturbance. Detailed descriptions on evaluation of these indices for eutrophication studies can be found in the work of Zardivar et al. (2008), Andersen et al. (1997).

There are a few indicators initially developed for lake waters (Vollenweider, 1969, 1976). Simple formulas based on relationships between phosphorus and chlorophyll concentrations have been proposed as eutrophication indices (Kirchner and Dillon, 1975). A Trophic State Index (TSI) has been proposed by Carlson (1977). This index combines three variables that include Secchi disk transparency, chlorophyll and total phosphorus. The assumptions on which TSI is based makes doubtful the use of the index at a worldwide scale.

Another potential tool in assessing trophic levels is the use of dominance curves. According to Clarke (1990) "*a dominance curve is based on a ranking of species (or higher taxa) in a sample in decreasing order of their abundance or total biomass*". The "*alternative to these dominance curves is to plot cumulative ranked abundances against log species rank as is so called* 'K-dominance' *curves*". If there is poor equitability among the species and the community is dominated by only a few species, then the curve is steep. Although the dominance curves have been used expensively in benthic studies (Lambshead et al., 1983; Clarke, 1990), the number of applications with phytoplankton data is rare. An application of K-dominance curves used by Ignatiades and Karydis (1990) using phytoplankton data from Saronicos

Gulf (Greece), did not pinpoint any eutrophic trends in the study area but simply showed seasonality. Many integrated indicators have been appropriate for detecting eutrophic trends. Most of them include oxygen concentration, water transparency and chlorophyll as effect variables whereas nutrients are the main pressure indicators (Baan and van Buuren, 2003).

The existence of numerous indicators can be perceived if we take into account: (a) different types of ecosystems require the use of specific indices as the importance of environmental variables (physical, chemical, biological and geological) varies from one ecosystem type to another and (b) some complex indices may require information useful in coastal management, policy making and marine spatial planning; in that case they need a wide spectrum of information.

2.5 Eutrophication Assessment: One-dimensional Statistics

2.5.1 *Descriptive statistics: the first step in assessing eutrophic trends*

It is the first step to "condense" filed data. The descriptors used most often are central tendency and variability. The mean is usually used for the central tendency and the standard deviation for the variability (Elliot, 1983). Minimum–maximum values are also used to define the limits of the distribution of a variable, the upper limit and the lower limit (Fowler et al., 1998). The application of summary statistics is the first step before any specific method of statistical analysis could be performed. However, it can provide information on the trophic status of an area if eutrophication scales based on nutrient, chlorophyll or cell number concentrations are available (Ignatiades, 2002a; Ignatiades et al., 1992; Stefanou et al., 2000). The Average Standard Deviation (ASD) method (Ebbesmeyer et al., 1991) is a simple and effective approach. This method was used to detect regime shifts during the period 1977–1989 in the North Pacific (Hare and Mantua, 2000). Stepwise description of the application is given by Mantua (2004).

Within the field of descriptive statistics is also a non-linear time series analysis method known as Recurrence Quantification Analysis (RQA) which has been proposed by Zardivar et al. (2008b) to identify regime shifts in environmental time series; this method is based on the Recurrence Plots (RP) and is characterized by some advantages regarding the assumptions of the method. It is independent of the data set size, data stationarity and the statistical distribution of the data. It can also be useful for a fast screen of the data sets but usually a combination with other techniques is advisable (Andersen et al., 2006b). Zardivar et al. (2008b) studied regime shifts in the Mediterranean coastal lagoon, Sacca di Goro, in Italy. The results indicated that for the identification of a single threshold in a time series RPs are robust even against high noise levels (up to 100%).

2.5.2 *Frequency distributions: probabilistic assessments*

Frequency distributions of variables related to water quality have been used by field workers for a long time ago (Heyman et al., 1984). In eutrophication studies, it was found that the relevant variables followed the log normal distribution although beta and gamma distribution patterns were also observed (Ignatiades, 2002a). If a researcher fits his data to a known frequency distribution, the distribution can then be used as a probabilistic system to assess trophic levels of an area under survey by using a small number of samples provided that standard distribution curves have been set (Bower and Kneese, 1968; OECD, 1984). The possibility of exceeding a critical value (may be a maximum acceptable concentration of a variable) can be assessed. Ignatiades et al. (1992) has developed standard curves of phosphate, nitrate, nitrite and ammonia of an area. A similar approach was also used by Giovanardi and Tromellini (1992) using chlorophyll concentrations. Stefanou et al. (2000) developed a non-parametric procedure to define the central tendency and dispersion of the distribution. This way distortion of the physical information of the data was avoided as no transformation for normality was applied. The standard distribution was named "Physical-Normal" (PN) distribution and formed the basis for assessing eutrophication on a probabilistic basis. If the distribution of the data is known, it is easy to apply data transformation to normalize the data distribution; it is then possible to apply parametric procedures for further data analysis.

2.5.3 *Outliers: identifying extreme eutrophic conditions*

The definition of an outlier has been given by Hawkings (1980): "an outlier is an observation which deviates so much from other observations as to arouse suspicions that it was generated by a different mechanism". The discussion on outliers goes back to 1777 when Bernoulli (1977) questioned the widespread practice among his contemporaries to discard discordant observations in the absence of *a priori* information. Seventy five years later, in 1852, Peirce was still wondering about the practice of rejecting discordant observations without any objective criterion: *"geometers have, therefore, been in the habit of rejecting those observations which appeared to them liable to unusual defects, although no exact criterion has been proposed to test and authorize such a procedure, and this delicate subject has been left to the arbitrary discretion of individual computers"* (Peirce, 1852). However, the simplest definition has been given by Grubbs (1969) who stated that "an outlier, is one that appears to deviate markedly from the other members of the sample in which it occurs". It was in 1878 that for the first time a criterion was proposed for rejecting discordant observations which was later called by Barnett and Lewis (1994) the bizarre criterion: *"the principle upon which it is proposed to solve the problem is that the proposed observations should be rejected when the probability of a system of errors obtained by retaining them is less than that of the system of errors obtained by their rejection multiplied by the probability of making so many, and no more, abnormal observations"*. This approach was the basis for developing objective ways to identify outliers. Today the term "outliers" is used by the investigators for any kind of observation that "stands apart from the bulk of the data that have been called "discordant observations", "rogue values", "contaminants", "surprising values", "mavericks" or "dirty data". Two mechanisms have been identified for the generation of outliers: outliers can be either the tail-values from a heavy tailed distribution, for example a log normal distribution or when the data actually arise from two different distributions. In this case there is the "main" or "basic" distribution and the "contaminants distribution" that is the distribution that produces contaminants. In the case of using field data, it is possible that an outlier may be the outcome of natural variation rather than any kind of weakness of the statistical (distribution) model used. It is therefore obvious that by rejecting outlying concentrations, environmentally important information, characterizing extreme conditions of pollution will be lost. Georgopoulos and Seinfeld (1982) working on variables characterizing air quality noticed the importance of outlying values in extreme atmospheric pollution episodes. This idea has been tested in eutrophication assessment studies with encouraging results. Data on phosphate concentrations collected in Saronicos Gulf, Greece, from nine stations covering the inner part of the Gulf receiving municipal effluents from the metropolitan area of Athens, were used for detection of outlying values. It was found that the number of outlying observations was increasing with the proximity of the sampling sites to the sewage outfall (Karydis and Ignatiades, 1982). Further work on outliers has been carried out using a multivariate approach (Karydis, 1994). Nitrate, ammonium, phosphate and chlα values from ten stations spread along the coastal area of the city of Rhodes, Greece, formed the set of data that was used for the detection of outlying values. The criterion was the "non-parametric box-and-whisker-plot" procedure. The mean values of outliers for each station were calculated and cluster analysis was applied. The tree diagram showed good discrimination among the oligotrophic, mesotrophic and eutrophic stations. However, the assessment of eutrophication levels by using outlying values has not received the attention it deserves from the researchers so far, in spite of the fact that there are well formulated procedures in literature (Barnett and Lewis, 1994; Beckman and Cook, 1983).

2.5.4 *Analysis of variance: assessing levels of eutrophication*

It is well known that measurements based on patterns and processes in the marine environment are surrounded by much "noise" that is high levels of variability in both time and space. The need for acquisition of accurate and reliable estimates has led the researchers to develop and elaborate experimental designs within the field known as "analysis of variance" (ANOVA) which can provide answers to different hypothesis sets (Underwood, 1981). The term ANOVA includes a family of statistical methods to compare the means of two or more variables or factors. Depending on the number of factors ANOVA can be one-way (simple ANOVA), two or even three-way ANOVA, nested design ANOVA, split-plot design and multifactor ANOVA (Sokal and Rohlf, 1981). However, extensive abuse of ANOVA in papers published in marine journals has been noticed. Underwood (1981) has identified plethora of problems in 151 papers he investigated, most of them concerning formulation of hypothesis presentation, computation or interpretation. Researchers seem to face difficulties with controls, independence of variables, replication and violation of the assumptions underlying ANOVA applications. More specifically, the model for the analysis was incorrect, multifactorial analysis was applied by simpler analyses leading to information losses due to interactions, violation of assumptions related to homogeneity were observed as well as data transformations that were applied inappropriately.

One-way analysis has been reported (Ignatiades, 1990) as a tool for testing the effects of depth (one factor) on nutrients and chlα (Ignatiades, 1995). Similarly, the effects of three factors namely season, sampling site and depth were tested using chlα concentrations in the SE Aegean Sea (Ignatiades et al., 1995). However, the potential of ANOVA *"sensu stricto"* in assessing eutrophic levels was investigated by Primpas and Karydis (2010a): the set of data (Ignatiades et al., 1981) was set up by nutrient and phytoplankton variables from nine stations in Saronicos Gulf. The data from the nine stations and each variable were tested using ANOVA as: (a) raw data (b) transformed data and (c) simulated data. After the application of the ANOVA, pairwise comparisons were carried out; it was found that the simulated data had the maximum sensitivity in discriminating differences among sampling sites (stations). ANOVA has also been applied to evaluate 20 phytoplankton indices and their efficiency in eutrophication studies (Tsirtsis and Karydis, 1998). The pairwise multiple comparison tests often used to compare polluted to control sites have been reviewed extensively by Jaccard et al. (1984) and Day and Quinn (1989). Emphasis has been laid on published work regarding violation of the ANOVA assumptions with special reference to heterogeneity and normality. In spite of the wide range of pairwise tests, the applications on eutrophication problems are very limited (Underwood, 1981; Underwood, 2007). It is not therefore possible to know the efficiency and shortcomings of these methods in assessing marine eutrophication levels. There is a limited number of papers on phytoplankton richness and abundance constrained by environmental variables in transitional waters in the Mediterranean Sea (Vadrucci et al., 2008). Spatial differences have been tested using abundance and species richness by the use of the two-way ANOVA. It was also found that the role of temperature was noticeable.

2.5.5 *Regression and correlation: interpreting causal-effect relationships*

Fairly often, the understanding of eutrophication processes requires the study of relation between two or more variables. In most cases the dynamics of a driving force, usually nutrients, with the outcome which may be excessive algal growth is under investigation. If a quantification of the relationship is needed, the method applied is fairly often the linear regression. The driving force is known in statistics as the independent variable whereas the variable describing the outcome (excessive plant biomass) is the dependent variable. If it is not clear which are the dependent/ independent variables or their physical connection is not known, then the method applied is known as correlation analysis.

Although there are many correlation coefficients (Sokal and Rohlf, 1981), the most commonly used is the product-moment correlation. There are many publications where linear correlation between nutrient variables (phosphate, nitrate, ammonia) and phytoplankton in terms of chlα, cell number, species richness or any other ecological indicator has been applied. In eutrophication studies when the dependent variable usually phytoplankton is correlated with physical and chemical variables, the result may be influenced by: (a) temperature and water transparency (b) nutrient concentrations and (c) random fluctuations from variables (even sometimes unknown) that have not been included in the statistical model.

A more specialized correlation method is the partial correlation. Partial correlation is applied when there is a group of variables but no distinction is possible between independent and dependent variables (Poole, 1974). Using this method of correlation, the contribution of each pair of variables can be assessed and the percentage of variation due to the rest of the variables can be quantified.

There are many publications using correlation in phytoplankton studies. On one hand they provide an insight regarding the understanding of system processes and dynamics but on the other hand the limitations of this statistical tool due to methodological assumptions are notable. Relevante and Gilmartin (1995) have studied the relationship among phytoplankton cell density, chlα and total carbon to understand aspects of the marine ecosystem dynamics in the Northern Adriatic. Using data from Saronicos Gulf, Ignatiades et al. (1987) applied partial correlation to understand the relationship between dark fixation and the ratio productivity/biomass. It was found that partial correlation coefficients were higher than simple correlation coefficients due to correction of the nutrient concentration and cell density variables of the matrix. The efficiency of correlation analysis was also studied in an oligotrophic system in the South Aegean (Vounatsou and Karydis, 1991): correlations between nutrients and chlα were rather poor; this may be due either to very low nutrient concentrations characterizing oligotrophy or to the contribution of other factors in those systems such as metallic elements and vitamins not measured in routine work. In addition, physical processes such as upwelling and oxidation of organic matter may add to the noise of the correlation matrix. The correlations between nutrients and phytoplankton variables were statistically significant in mesotrophic and eutrophic waters (Ignatiades et al., 1985).

More information is acquired by regression models, provided that it is well established which variable is the dependent variable. In regression applications, the simple regression, the multiple regression and the stepwise regression models can be used. These methods have been described by Draper and Smith (1998). Although the linear regression is very popular among the marine researchers, including eutrophication and planktonic studies, mainly to describe the effects of nutrient and physical variables of phytoplankton, there is doubt about examining the assumptions of the statistical model before proceeding to the analysis of the data (Poole, 1974). Estrada et al. (1993) has used multiple regression to study the dependence of primary production on chlorophyll, water column stability and depth. The main problem when multiple regression is applied is the presence of interactions among independent variables (Poole, 1974). Each interaction between variables accounts for an independent variable and therefore the model becomes complicated. In addition, the number of observations per variable decreases.

The stepwise regression seems to be a more versatile and interactive statistical tool. The independent variables (X_i) are ranked according to their proportion of the corrected sum of squares of the dependent variable (Y) they explain. A paradigm using stepwise regression has been given by Vounatsou and Karydis (1991). The dependent variable (chlα) was regressed over four independent variables that is phosphate, silicate, nitrate and ammonia. The variables were transformed for normality: reciprocal and exponential transformations were found as the most efficient for good model fitting. The concentration of chlorophylls as a function of nitrates in Thermaikos Gulf, a eutrophic area in the Northern Aegean, was investigated using a piecewise linear regression (Nikolaidis et al., 2006).

Bloom indicators were correlated with nutrient concentrations over a period of a few years in an effort to understand regime shifts and assess threshold nutrient levels including harmful algal events along the European coastal waters (Stolte and Granelli, 2006). Stolte and Granelli have also studied phytoplankton trends using linear regression. The availability of large ecological time series allowed Mantua (2004) to apply Autoregressive Moving Average (ARMA) models as well as vector Autoregressive (VAR) for detecting regime shifts. There is a shortcoming in this process that random fluctuations can be confused as thresholds. By setting a statistical criterion at the 5% level at a predetermined value, the possibility of an error is eliminated. Objective means provided by the Vector Autoregressive (VAR) modeling helps to identify and assess the statistical significance of regime shifts (Mantua, 2004).

Autocorrelation is fairly often referred to as serial correlation between the data points of variables that are based on associated aspects. It has been reported that "*autocorrelation is a characteristic of data in which the correlation between the values of the same variables is based on related objects*" (Kitsiou and Karydis, 2011). A detailed description of these methods has been given by Pagou and Ignatiades (1988) and have been applied by the authors to detect seasonality patterns of phytoplankton along marine eutrophic coastal waters. Daily patterns of the dynamics and ecological significance of phosphorus internal load was studied using autocorrelation analysis (Istvanovics et al., 2004). The researchers found out that daily sampling was required for proper determination of the suspended particulate matter (SPM) as well as the coefficient of vertical light attenuation; this indicates that routing monitoring is not suitable to get a good insight into the environmental variability regarding phytoplankton dynamics in certain habitats such as shallow lakes. Spatial autocorrelation is discussed in the Section 2.7 "Spatial Analysis Methods and Mapping in Marine Eutrophication Studies".

2.6 Multi-dimensional Statistical Analysis

2.6.1 *General background*

Multivariate statistical methods used for assessing eutrophication involve a relatively small number of variables mainly nutrient concentrations, phytoplankton cell number, chlα, ecological indices and macrophyte biomass or community parameters. In addition, it is possible to add some physical variables such as temperature and water transparency that could help understand phytoplankton dynamics. Preliminary tests are applied on the raw data (Figure 2.1) to test whether the assumptions of the statistical procedures that follow are fulfilled. If not, some data manipulations such as variable transformation may satisfy the statistical requirements. In the present section, applications of multivariate procedures will be displayed regarding eutrophication assessment, detection of eutrophic trends as well as ranking of sampling sites according to their eutrophic state. These results are the "end products"

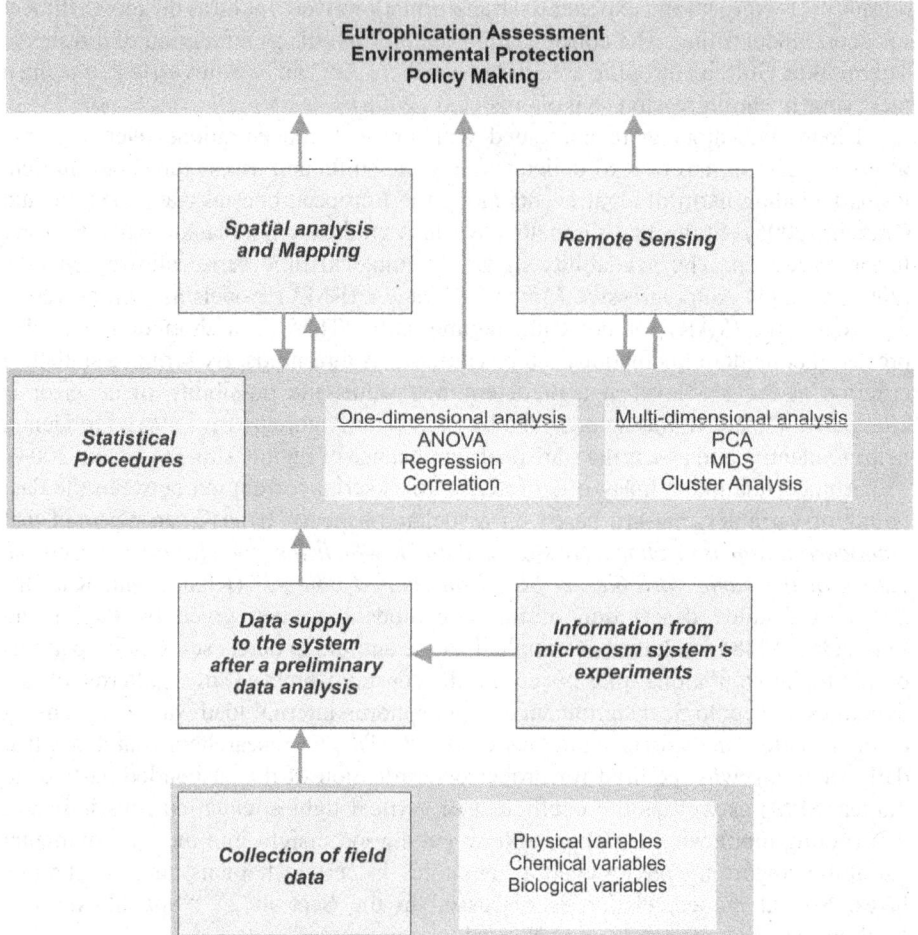

Figure 2.1 Flow chart showing the contribution of the statistical methodology for quantitative assessment of eutrophication, Environmental Protection and Policy Making.

in environmental studies. However, as more integrated approaches are required these days, the outcome of a multivariate procedure can be assimilated in a broader processing including socio-economic variables such as Multiple Criteria Analysis (MCA) methods and Spatial Analysis. This complex body of variables can be useful in coastal management (Agardy, 2010), environmental protection and Marine Spatial Planning (Portman, 2016; Kidd et al., 2011). It can also form a useful background for setting standards in environmental legislation such as national legislation, EU Directives and International Conventions. Kitsiou and Karydis (2011) provide information on the objectives, potential and shortcomings of multivariate methodology in marine eutrophication studies. Most of these methods have been characterized as ordination methods, a term that denotes the arrangement of units in an orderly way (Goodall, 1954; Sharma, 1996; Whittaker, 1982). The term originates from the Latin word "ordination" that means orderliness, organization, arrangement. The idea of these methods is to reorganize the information on a quantitative basis by reducing the voluminous sets of data and showing ecosystem trends and relationships between the ecosystems' components. The most popular ordination methods among ecologists include Principal Component Analysis (PCA), Multidimensional Scaling and Factor Analysis (Legendre and Legendre, 2003). These methods together with classification methods are the main multidimensional tools in marine ecology studies.

2.6.2 *Principal component analysis: a primer in reducing dimensionality*

The Principal Component Analysis or PCA is the oldest among the indirect ordination methods. Suppose that the data are available for n units such as phytoplankton species of nutrient variables, pigment variables, temperature, water transparency and p variables (sampling stations). The full set of data is then a matrix X, nxp dimensions, having n rows and p columns. The idea of the analysis is to produce an ordination of units in a reduced number of dimensions. The outcome of the procedure accentuates the major patterns of variation and their responses (Digby and Kempton, 1987; Dunteman, 1989). The data compression into a new space of lower dimensions is known as "dimensional reduction". This method is usually used at the present as an exploratory data analysis method; an example in river water quality assessment and quantification of anthropogenic influences has been presented (Vega et al., 1998). The variables that formed the data matrix were nutrients, oxygen concentrations, BOD and COD. However, publications focusing on the assessment of eutrophication are very limited. A work on the use of PCA for developing a multimetric index for eutrophication assessment has already been presented in a previous section (Primpas et al., 2010).

2.6.3 *Multidimensional scaling: a non-metric multivariate tool*

In Multidimensional Scaling (MDS), a triangular table of distances is constructed that forms the basis for presenting the relationships between many objects in two dimensions (Manly, 2001). This method became popular since the 80s in studies regarding multispecies patterns as well as in benthic marine work (Warwick et al., 1988). There are two advantages regarding applications of MDS: (a) it is a non-metric procedure and (b) it is accompanied by an ANOSIM test that is analysis of similarities; this test provides to the researcher additional information regarding statistical significance. However, the number of applications in eutrophication studies is limited (Karydis, 1992; Moriki and Karydis, 1994).

2.6.4 *Discriminant Analysis (DA)*

This method is used in marine ecology to classify objects into groups. This multivariate procedure is characterized by clarity of interpretation (Stevens, 1996). A eutrophication assessment classification scheme using DA has been proposed by Tsirtsis and Karydis (1999). Data sets collected from 14 sampling sites of known trophic states in the Aegean Sea and nitrate (V_1), phosphate (V_2) and chlorophyll (V_3) variables were used. These variables were used for the estimation of the coefficients of the two discriminant values:

$$\text{Function } 1 = 3.242 - 0.130 \times V_1 + 0.896 \times V_2 + 5.584 \times V_3 \tag{3}$$

$$\text{Function } 2 = 1.081 - 0.378 \times V_1 + 2.932 \times V_2 - 3.211 \times V_3 \tag{4}$$

where V_1, V_2 and V_3 are the logarithms of nitrate, phosphate and chlorophyll α concentrations. The two discriminant functions were used for plotting a map divided into three areas characterized as eutrophic, mesotrophic and oligotrophic respectively (Figure 2.2). This way, nutrient concentration from any sample collected from an area of unknown trophic status can be placed on the plot and the trophic status can be assessed.

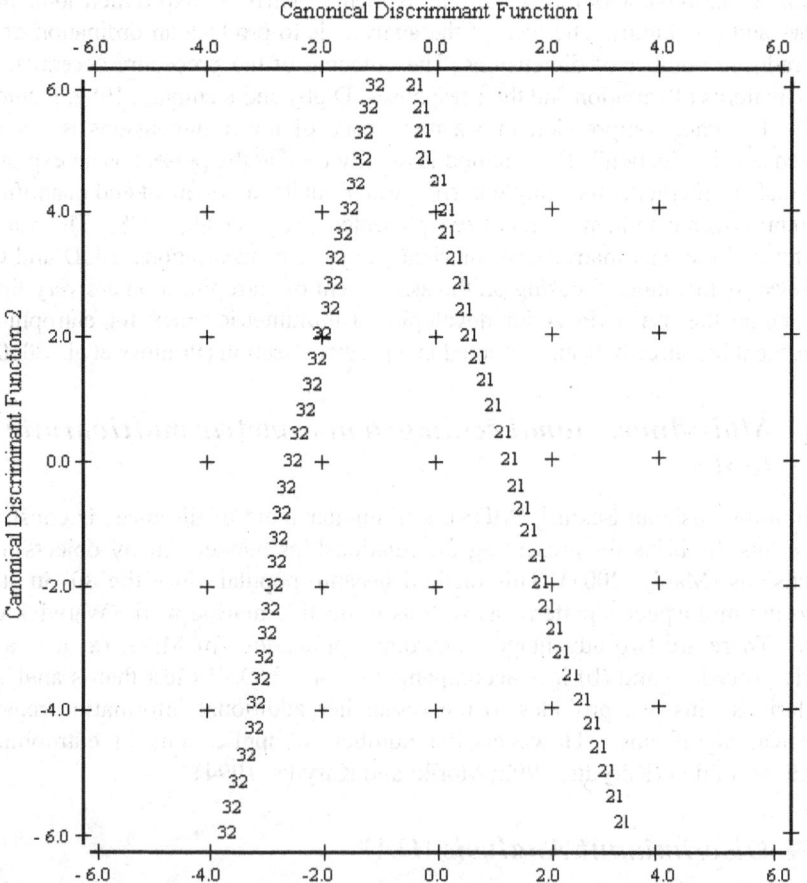

Figure 2.2 A discriminant plot for evaluation of the trophic status of marine coastal waters is shown. Eutrophic, mesotrophic and oligotrophic areas are shown in the plot demarcated by 1's, 2's and 3's symbols respectively. Stations under assessment are also shown in the designated areas (source: Tsirtsis and Karydis, 1999).

2.6.5 Correspondence Analysis (CoA)

Correspondence Analysis is considered among the most popular ordination methods among ecologists (Digby and Kempton, 1987). The method examines the relationship between stations and environmental variables, usually species, by simultaneous ordination (Pielou, 1984; Ludwig and Reynolds, 1988). Although there is published work on phytoplankton ecology (Ignatiades et al., 1996; Okolodkov and Dodge, 1996), applications on eutrophication assessment have not been reported.

2.6.6 Numerical classification

The main objective of numerical classification or Cluster Analysis (CA) used in ecological applications is to classify sites, environmental variables or species. According to van Tongeren (1987): "Cluster analysis is an explicit way of identifying groups in raw data and helps us to find structure in the data". As there is a continuity in nature, cluster analysis is actually partitioning the physical information into discontinuous subsystems known as clusters. In addition, cluster analysis provides the researcher with information on species concurrence (internal data structure), establishes community types in descriptive studies (syntaxonomy) and mapping–detecting relations between biological communities and environmental variables (external analysis). Besides, community analysis classification (non-hierarchical) can also handle redundancy, noise and outliers.

The quantification of the relationship between sampling stations and variables is the first step in cluster analysis of ecological data (Legendre and Legendre, 2003). The step requires the choice of a similarity (or distance) index. The second step is the allocation of the objects into groups or clusters according to their similarity. These clusters are usually presented in the form of a dendrogram (tree diagram). There is extensive literature on clustering techniques (Pielou, 1984; Gauch, 1989; Sharma, 1996; Everitt et al., 2001; Legendre and Legendre, 2003; Romesburg, 2004). However, the efficiency of numerical classification in ecological and environmental applications is debatable (Cao et al., 1997). This is mainly due to three reasons: (a) the criteria of selecting similarity or distance measures are not always clear, depending sometimes on the personal choice of the researcher. The selected algorithm influences the outcome of the method (Ludwig and Reynolds, 1988; Pielou, 1984), (b) the choice of clustering methods influences to a great deal the structure of the dendrogram (Everitt et al., 2001; Primpas et al., 2008) and (c) there is often uncertainty regarding the "structure" of the ecological/environmental body of data.

The choice of similarity measures is closely connected with bias to the outcome (Cao et al., 1997). This is why choosing a similarity measure, the nature of the data and the particular objective that has been set by the researcher has to be taken into account (Gower and Legendre, 1986). Distance coefficients based on Euclidean measures (Ludwig and Reynolds, 1988; Romesburg, 2004) seem to suit abiotic variables connected to eutrophication; these variables can be temperature, water transparency, nutrient concentrations, chlα values, phytoplankton cell number or even ecological indices and water quality indicators. It has been found that the Euclidean Distance overweighs dominant species even if the variables have been transformed (Cao et al., 1997). The same problem has been reported by Karydis (1992) in the eutrophication assessing procedure using nutrient and chlα concentration variables. On the contrary, the absolute distance $d_{(j,k)} = \Sigma_i |x_{ij} - x_{ik}|$ has provided the best resolution among clusters within the mesotrophic range; this is due to the fact that differences in nutrient and chlα variables were not squared: they were simply summed up. Another similarity measure, the Pearson Product-Moment correlation coefficient (Romesburg, 2004) has shown limited resolution in forming discrete clusters in sets of data (Everitt, 1981). In community analysis, Canberra metric coefficient and Bray-Curtis coefficient seem to be both efficient. Among these two coefficients, most popular is the Bray-Curtis index (Magurran, 2004). The idea of introducing species weights in the similarity matrix has added an advantage in eutrophication studies: as weights indicate extra significance for a species in connection with some particular characteristics, species weighing can be used for toxic microalgae in eutrophication studies using classification methods. When a clustering algorithm is chosen, the principal criterion should be discriminant efficiency of the algorithm, that is, its efficiency to distinguish among oligotrophic, mesotrophic and eutrophic sampling sites. Cao et al. (1997) working on river invertebrates, tested four clustering algorithms: TWINSPAN, Ward Linkage, Compete Linkage and UPGMA (Average Linkage Clustering Method). The most efficient agglomerative algorithm in discriminating tributaries was the Ward Linkage. The second discriminant power was the TWINSPAN algorithm. In a study on eutrophication assessment (Primpas et al., 2008), eight clustering algorithms were used to evaluate their efficiency in discriminating among oligotrophic, mesotrophic and eutrophic sampling areas. The best resolution was shown by the "Ward's" clustering algorithm. Another criterion for choosing a clustering algorithm is connected with the stability of the clustering structure. A clustering structure is considered stable if minor changes in the species matrix do not modify substantially the clustering pattern (Hajdu, 1981). Cluster analysis combined with other multivariate methods has been applied by Ignatiades (2002b) on nutrient and phytoplankton data. However, there was no statistical significance on the results. The need for exploratory work in cluster analysis has been stressed a long time ago (Boelsch, 1977). Unfortunately, there are a limited number of publications regarding criteria selection and suitability of both similarity indices and clustering algorithms. An advancement in cluster analysis has been proposed by Vassiliou et al. (1989). An algorithm was developed which generated simultaneous permutations on two or three columns of the matrix at a level of significance (Everitt et al., 2001). Although this algorithmic procedure has not been used widely in ecological applications, Giancarlo and Ultro (2012) have evaluated by using paradigms, the statistical significance of this algorithm for cluster validity.

2.7 Spatial Analysis Methods and Mapping in Marine Eutrophication Studies

In marine eutrophication studies, the assessment of the spatial structure and the heterogeneity in data is carried out by application of spatial analysis methods.

Heterogeneity refers to variations in the measurements performed during sampling at discrete sampling sites. Structure functions are used to describe spatial structures allowing the quantification of the spatial dependency and its partitioning among distance classes (Legendre and Legendre, 2003). The most commonly used structure functions are correlograms and variograms.

The spatial correlogram is a graph with the autocorrelation values plotted in y-axis against distance classes among sampling sites. The Moran's I or Geary's c spatial autocorrelation statistics can be calculated to measure spatial autocorrelation (Cliff and Ord, 1973). The variogram is the basic tool of geostatistics and provides the means for the detection of anisotropies and the calculation of the degree of homogeneity in data sets (Mabit and Bernard, 2007). It is expressed as a graph with the semi-variance $\gamma(h)$ plotted in y-axis against distance classes among sampling sites (h, lag). When anisotropy is present in data, the autocorrelation function is not the same for any geographic direction. The maximum value of $\gamma(h)$ is called sill and indicates the absence of spatial dependence in data. The semi-variance value $\gamma(h)$ when h = 0 is the nugget variance and represents the local variation occurring at scales finer than the sampling interval as well as the measurement and sampling error.

Directional correlograms and variograms can be used to assess the spatial structure of variables at specific directions. The determination of the directions where different degrees of anisotropy are observed is possible using the surface variogram (Michelakaki and Kitsiou, 2005).

In two-dimensional space, the spatial structure and spatial distribution of variables can be illustrated on thematic maps. When irregularly spaced point measurements collected during sampling surveys are available, interpolation methods provide the means for the conversion of such fragmented information into a surface with smooth gradient of data values (Burrough and McDonnell, 2000). A wide range of interpolation methods for creating such surfaces is available; they are grouped in two general categories, global and local. Regarding the global ones, a single function describing a surface is calculated covering the entire map area. On the other hand, the local methods estimate the surface at successive points using only the selected nearest data points (Li and Revesz, 2004). In marine eutrophication studies, the most frequently used interpolation methods are: (i) The Trend surface analysis or trend surface mapping, a global surface-fitting procedure (Lancaster and Salkauskas, 1986), (ii) the Inverse Distance Weighted (IDW) methods, where the interpolating surface is a weighted average of the scatter points with the weight assigned to each scatter point diminishing as the distance from the interpolation point increases (Weber and Englund, 1994), (iii) the Kriging, a local interpolation method based on the assumption that the variable to be interpolated is considered as a localized variable (Matheron, 1963). The calculation and analysis of variograms is a pre-requisite for the estimation of the spatial structure of variables and the calculation of the optimum weights at different sampling distances (Cressie, 1993).

Spatial analysis methods have been widely applied for the assessment of marine eutrophication. It should be noted that the majority of these approaches have been implemented in the framework of Geographical Information Systems (GIS), since GIS provide the appropriate environment for the management, processing and analysis of spatial data sets and the development of thematic maps.

According to Dutilleul (1993) and Legendre (1993), a variable can be characterised as homogeneous or heterogeneous in space regarding the scale at which it is studied; therefore, they used spatial autocorrelation calculations for the planning of ecological experiments. Autocorrelation analysis has been also applied by Blanchet et al. (2008) on a data set derived from SeaWiFS satellite data in order to assess the heterogeneity of the spatial distribution of chlorophyll concentration seasonal dynamics.

Anttila et al. (2008) developed an optimal sampling network by applying geostatistics to investigate patchiness of water quality related to sampling methodology. Another attempt to develop an optimal sampling design for eutrophication studies has been proposed by Kitsiou et al. (2001); 34 sampling sites of chlorophyll concentrations located in a grid pattern of 1 × 1 NM were reallocated based on optimized sampling strategy and the calculated anisotropy using variograms. Eleveld and van der Woerd (2006) investigated the variability of chlorophyll and suspended particulate matter concentrations derived from SeaWiFS satellite data in order to assess the range where the measurements become independent by performing semi-variogram analysis. A methodology including calculation of semivariance and discrete transinformation entropy has been applied to calculate the optimum range of the monitoring distance for phosphorus and nitrogen which are the most important nutrients that lead to eutrophic conditions. The study indicated the ranges of 28 to 82 and 37 to 50 km for the phosphate and nitrate + nitrite, respectively in San Francisco bay, USA (Boroumand et al., 2018).

The IDW method was applied by Newton et al. (2003) to interpolate data sets and produce maps of a number of parameters related to eutrophication in the coastal lagoon of Ria Formosa, Portugal. Eutrophication risk assessment of small water bodies was carried out by Kovacs and Honti (2008) using a GIS-based methodology for the estimation of phosphorus emissions at small scale watersheds. Kovač et al. (2014) applied a methodology based on self-organizing maps to classify the pigment structures of the Adriatic Sea into a number of patterns present in the collected data. Furthermore, the severe eutrophication status and spatial trend in the coastal waters of Zhejiang province (China) was assessed by estimating the spatial distributions of chemical oxygen demand, dissolved inorganic nitrogen, and dissolved inorganic phosphorus across space by application of a novel method based on the Bayesian maximum entropy (BME) which gave better results compared to kriging and IDW techniques (Jiang et al., 2018).

Marine eutrophication assessment requires the study of a high number of parameters with different spatial structure. Consequently, co-estimation of their spatial patterns is needed for the accurate estimation of existing eutrophication trends (Valiela, 1995). The synthesis of different spatial patterns can be performed by application of various approaches. Kitsiou and Karydis (1998) used multiple criteria choice methods to synthesize the spatial distributions of nutrients and phytoplankton cell number to produce the eutrophication map of Saronicos Gulf, Greece. In the same area, the spatial distributions of seven ecological indices were integrated using the GIS overlay technique (Kitsiou and Karydis, 2000). The same technique was also applied by Xu et al. (2001) to synthesize the spatial distributions of six chemical, physical and biological indicators of eutrophication. Kong et al. (2017) proposed a methodology for real-time eutrophication status evaluation of coastal waters using support vector machine with grid search algorithm and co-estimation of the six parameters related to eutrophication.

Finally, it should be mentioned that the accuracy of the spatial distributions related to eutrophication variables which are produced by interpolation, depends on the accuracy of the data set collected during sampling surveys and the effectiveness of the selected interpolation method. Therefore, when spatial analysis methods and in particular interpolation methods are applied, the sampling sites network should be optimized based on the extent and previous knowledge of the conditions in the study area as well as the spatial behavior of the variables under study. Furthermore, the effectiveness of the performance of interpolation methods can be assessed by use of cross-validation and the calculation of specific statistics (Webster and Oliver, 2001).

2.8 Use of Remote Sensing in Assessing Eutrophic Conditions

During the last couple of decades, remote sensing has proved valuable to assess eutrophication levels in marine areas and analysis of the observed trends, especially at large spatial scales. Ocean color, a unique property measured from satellite sensors provides information from the sea surface to a few tens of meters depth (Maritorena and Siegel, 2005). Regarding marine eutrophication, the abundance and distribution of phytoplankton chlorophyll in seawater is estimated providing surface patterns of phytoplankton biomass distribution at large spatial scales (Garcia et al., 2005).

These days, a wide range of data sets from various satellite sensors are available. CZCS was the first instrument to measure ocean color, especially chlorophyll concentrations (Iverson et al., 2000). Landsat MSS/TM/ETM+ and SPOT sensors have been also successfully used to assess marine eutrophication (Torbick et al., 2008). The launch of SeaWiFS in 1997 and later of MODIS/AQUA and ENVISAT (2002) with MERIS instrument made available large sets of ocean color data (Werdell et al., 2009; Bresciani et al., 2014). The recent Sentinel missions support marine monitoring by providing timely, continuous and independent data regarding the behavior, use, and health of the oceans and coastal zones (http://marine.copernicus.eu).

Satellite data in different levels of processing can be retrieved from a number of available portals, i.e., EMODnet (European Marine Observation and Data Network), the Copernicus Marine Environment Monitoring Service, NASA Ocean Color (Goddard Space Flight Center, Ocean Biology Processing Group (OBPG)), etc. In literature, an extended number of approaches are available where satellite data is used either to assess the eutrophication status in various study areas based on already developed algorithms or to be combined with *in situ* measurements to produce new or optimize existing algorithms.

An overview of the use of remote sensing to assess algae blooms is available by Klemans (2012). Gohin et al. (2008) classified the coastal waters of the French Atlantic continental shelf and the English Channel using SeaWiFS satellite data based on the eutrophication risk criterion of the Water Framework Directive (WFD). Harvey et al. (2015) compared the chlorophyll-α concentrations retrieved from MERIS data with ship-based monitoring for the productive seasons in a coastal area of the Baltic Sea. The comparisons proved the reliability of the satellite-based monitoring system, since the estimations of chlorophyll-α concentration were comparable to *in situ* measurements in terms of accuracy. Spatio-temporal variability of sea surface temperature (SST) and chlorophyll was evaluated using MODIS products from 2002 to 2013 in the Persian Gulf by analysis of the spatio-temporal stability and abnormality of MODIS SST and chlorophyll (Moradi and Kabiri, 2015). Colella et al. (2016) computed chlorophyll trends over the Mediterranean Sea by using satellite data, and documented the efficiency and reliability of remote sensing to control the "good environmental status" (i.e., the Marine Directive of EU marine waters by 2020) and to support the application of international regulations and environmental directives.

In conclusion, satellite data are valuable in marine eutrophication studies, especially at large spatial scales and in particular when analysis of existing trends is required. Their effectiveness increases when they are combined with *in situ* measurements in order to develop appropriate algorithms for the calculation of concentrations of representative variables. The latter is due to a number of limitations related to the acquiring data sets from space. First of all, remote sensing can operate only on the sea surface skin, since electromagnetic radiation is very poor at penetrating water. Secondly, electromagnetic radiation at specific wavelengths cannot penetrate clouds, therefore frequently there is lack of data from extended areas and third, the temporal resolution of the majority of satellite imagery is limited. Furthermore, during marine eutrophication assessment at low spatial scales, satellite data of very high spatial resolution are not often available and supplementary *in situ* measurements are required. Finally, the selection criteria of the most appropriate satellite data sets for assessing marine eutrophication are the technical characteristics of the satellite sensor, i.e., spectral, spatial, temporal resolution, the spatial scale and the morphological characteristics of the study area.

2.9 Use of Multiple Criteria Analysis in Assessing Eutrophication

Multiple Criteria Analysis (MCA) methods are used for ranking a number of alternative choice possibilities/hypotheses, based on a number of criteria (Voogd, 1983; Nijkamp and Voogd, 1986), which can be expressed in both metric and non-metric form. In addition, weights can be assigned to criteria, in cardinal or ordinal units, to highlight their relative importance.

A wide spectrum of MCA methods are available (Malczewski, 2006) and all of them obey the principle of the pairwise comparison of the scores for all the alternatives and for each criterion (Nijkamp and Voogd, 1986). On the other hand, a number of weighting algorithms are also available, ranging from simple direct methods to more complex approaches. The first step to apply any MCA method is the creation of the Impact Matrix which has the following form, when, i hypotheses and j criteria are considered.

$$\text{Impact Matrix} = \begin{vmatrix} b_{11} & \cdots & b_{1j} \\ b_{21} & \cdots & b_{2j} \\ b_{i1} & \cdots & b_{ij} \end{vmatrix} \tag{5}$$

where b_{ij} is the score of the hypothesis i according to the criterion j. In the case that $b_{1j} > b_{2j}$, the hypothesis I_1 dominates over the I_2 as far as the j criterion is concerned (Hartog et al., 1989).

The next step is standardization of the Impact Matrix and assignment of weights to the criteria according to the selected weighting algorithm. Finally, application of the selected scoring algorithm to rank the alternative choice possibilities.

Among the numerous MCA methods available in literature, the Weighted Summation is the most popular due to its simplicity and reliability of the results (Howard, 1991). The Evamix method (Voogd, 1983) makes a separation of the quantitative (cardinal) from the qualitative (ordinal) data in the Impact Matrix, before the application of appropriate algorithms for each category (Hajkowicz and Higgins, 2008). The ELECTRE methods are widely known (i.e., ELECTRE II, III and IV) as concordance analysis which focuses on the degree to which one alternative performs better than another (Govindan and Jepsen, 2016). Both an index of concordance and an index of discordance are used to create a dominance relationship for each pair of alternatives (Wang et al., 2009). The Regime method is an ordinal generalization of pairwise comparison methods calculating the probability that one alternative ranks above another (Janssen, 2001).

MCA methods have been applied to assess marine eutrophication trends as well as to support coastal management and decision-making (Hajkowicz, 2007; Ananda and Herath, 2009). Regime method has been applied in assessing coastal marine eutrophication problems (Moriki and Karydis, 1994) and proved effective for the discrimination of eutrophication levels in coastal areas.

Almasri and Kaluarachchi (2005) presented an integrated methodology for the optimal management of nitrate contamination of ground water by combination of environmental assessment and economic cost evaluation. A MCA application for mapping the risks of agricultural pollution for water resources has been proposed by Giupponi et al. (1999) by co-estimating soil and climate variables and alternative land uses. Koo and O'Connell (2006) applied an integrated modeling and multi-criteria analysis to evaluate a set of land use alternatives and to identify an 'ideal' compromise between economic return and environmental pollution. In addition, Schuwirth et al. (2018) applied multi-criteria decision analysis for integrated water quality assessment and management support by combining multi-criteria decision support methods with water quality modelling and scenario planning. Finally, Pang et al. (2017) applied multi-criteria decision analysis for the management of Harmful Algal Blooms (HABs) by analysing the priorities of various stakeholders regarding HAB mitigation.

Case study

A case study where MCA has been applied to assess the quality of the marine environment is presented below (Spitieri, 2017). The study area is Strymonikos Gulf in the North Aegean Sea, Greece. The rivers Strymonas and Richios flow out in the Gulf, while sea water from the Black Sea arrives there as well (Sylaios et al., 2006). In the coastal area, there are a number of settlements and a wide range of touristic activities are carried out, since there is a long sandy beach which is visited by many tourists during the summer period. A sampling survey was carried out in September 2011 (Kitsiou et al., 2011) to collect data to assess the eutrophication conditions of the Gulf. A network of 17 sampling sites was designed and water samples collected from 1 m depth. Concentration of chl-a (μg/l), N-NO$_3$ (μM) and P-PO$_4$ (μM) were measured and stored in a spatial database in a GIS environment. Using spatial analysis, the study area was divided in 17 sub-zones. In particular, Thiessen polygons (Mu 2009) were developed based on the sampling sites network. Each sub-zone corresponded to a sampling station and every point in the sub-zone i was located in a distance from the sampling site i lower than from any other sampling station. The concentrations of the measured variables in each station were considered representative of the relative sub-zone. Therefore, the 17 sub-zones were ranked based on their eutrophication status by application of MCA analysis; in the Impact matrix, the sub-zones were the alternative choice possibilities and the concentrations of chl-a (μg/l), N-NO$_3$ (μM) and P-PO$_4$ (μM) the criteria. The Regime MCA method was used where the assignment of weights is qualitative, performed by ranking the criteria based on their importance (Janssen, 2001). As already mentioned, the result of the Regime method was the ranking of the sub-zones according to their eutrophication status and not their classification to specific eutrophication levels. In order the latter to be possible, three (3) more alternatives, C1, C2, C3 were added in the Impact matrix, which represented the boundary conditions of the eutrophication levels (Table 2.1). The values of the criteria of these new alternatives/sub-zones are shown in Table 2.2.

Table 2.1 Eutrophication scales for chl-a (μg/l), N-NO$_3$ (μM) and P-PO$_4$ (μM) (Simboura et al., 2005; Kitsiou et al., 2002).

chl-a (μg/L)	0.00		0.100		0.600		2.210	
N-NO$_3$ (μM)	0.00	Oligotrophic	0.620	Lower mesotrophic	0.650	Higher/Upper mesotrophic	1.190	Eutrophic
P-PO$_4$ (μM)	0.00		0.070		0.140		0.680	

Table 2.2 The criteria values of the additional alternatives representing the boundary conditions of the eutrophication levels (O: Oligotrophic, LM: Lower mesotrophic, UM: Higher/Upper mesotrophic, E: Eutrophic).

	C1 O -> LM	C2 LM -> UM	C3 UM -> E
chla (μg/L)	0.100	0.600	2.210
N-NO$_3$ (μM)	0.620	0.650	1.190
P-PO$_4$ (μM)	0.070	0.140	0.680

The Regime method was applied twice: (i) without assigning weights to the criteria and (ii) by ranking the importance of the criteria (chlα > $N-NO_3$ > $P-PO_4$).

The results of (i) are shown in Table 2.3 and illustrated on the thematic map produced (Figure 2.3).

According to the results, the majority of the sub-zones of the Gulf was characterized as oligotrophic with only the sub-zones 2, 3 and 6 characterized as lower mesotrophic. These sub-zones are near the estuaries of Strymonas and Richios rivers as well as residential areas.

The results of (ii) are shown in Table 2.4 and illustrated on the thematic map produced (Figure 2.4).

In this case, the sub-zones 6, 7, 3 and 17 were characterized as lower mesotrophic, while all the others as oligotrophic. Compared to case (i), the sub-zones 7 and 17 are here lower mesotrophic than oligotrophic and the sub-zone 2 has been classified as oligotrophic.

The results indicate that the combination of spatial analysis methods and MCA analysis can provide a useful tool to assess marine eutrophication. The methodology described allowed the classification of the sub-zones to the four eutrophication levels. In addition, information on their relative rank within each eutrophication level is provided. Furthermore, an important point is the co-estimation of three variables related to eutrophication; as it is widely known, eutrophication is a multi-parametric phenomenon and methodologies based on the integration of the information from

Table 2.3 Ranking of the sub-zones without assigning weights to the criteria.

Ranking	Sub-zone	Eutrophication level
1	C3	**Higher/Upper mesotrophic -> Eutrophic**
2	C2	**Lower mesotrophic -> Higher/Upper mesotrophic**
3	ZONE 3	Lower mesotrophic
4	ZONE 6	Lower mesotrophic
5	ZONE 2	Lower mesotrophic
6	C1	**Oligotrophic -> Lower mesotrophic**
7	ZONE 16	Oligotrophic
8	ZONE 17	Oligotrophic
9	ZONE 7	Oligotrophic
10	ZONE 1	Oligotrophic
11	ZONE 12	Oligotrophic
12	ZONE 15	Oligotrophic
13	ZONE 14	Oligotrophic
14	ZONE 5	Oligotrophic
15	ZONE 10	Oligotrophic
16	ZONE 11	Oligotrophic
17	ZONE 8	Oligotrophic
18	ZONE 9	Oligotrophic
19	ZONE 13	Oligotrophic
20	ZONE 4	Oligotrophic

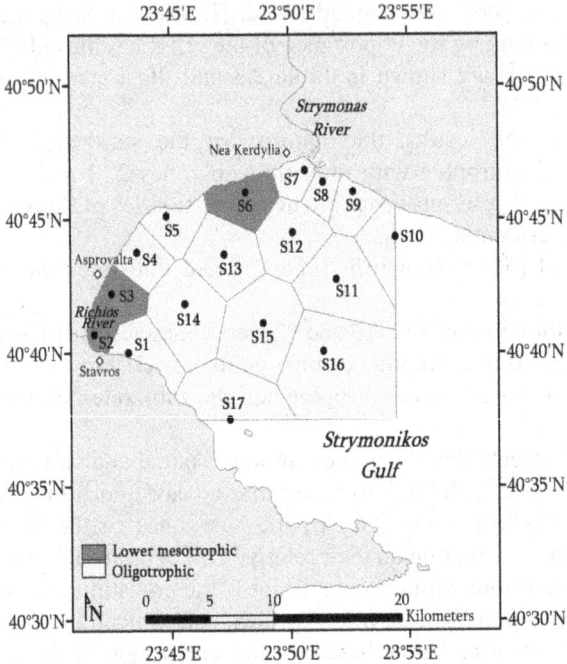

Figure 2.3 Thematic map of eutrophication levels in Strymonikos Gulf by application of the Regime MCA method without assigning weights to the criteria.

Table 2.4 Ranking of the sub-zones by assigning weights to the criteria (chlα > N-NO$_3$ > P-PO$_4$).

Ranking	Sub-zone	Eutrophication level
1	**C3**	**Higher/Upper mesotrophic -> Eutrophic**
2	**C2**	**Lower mesotrophic -> Higher/Upper mesotrophic**
3	ZONE 6	Lower mesotrophic
4	ZONE 7	Lower mesotrophic
5	ZONE 3	Lower mesotrophic
6	ZONE 17	Lower mesotrophic
7	**C1**	**Oligotrophic -> Lower mesotrophic**
8	ZONE 2	Oligotrophic
9	ZONE 16	Oligotrophic
10	ZONE 8	Oligotrophic
11	ZONE 12	Oligotrophic
12	ZONE 1	Oligotrophic
13	ZONE 5	Oligotrophic
14	ZONE 9	Oligotrophic
15	ZONE 10	Oligotrophic
16	ZONE 11	Oligotrophic
17	ZONE 15	Oligotrophic
18	ZONE 14	Oligotrophic
19	ZONE 4	Oligotrophic
20	ZONE 13	Oligotrophic

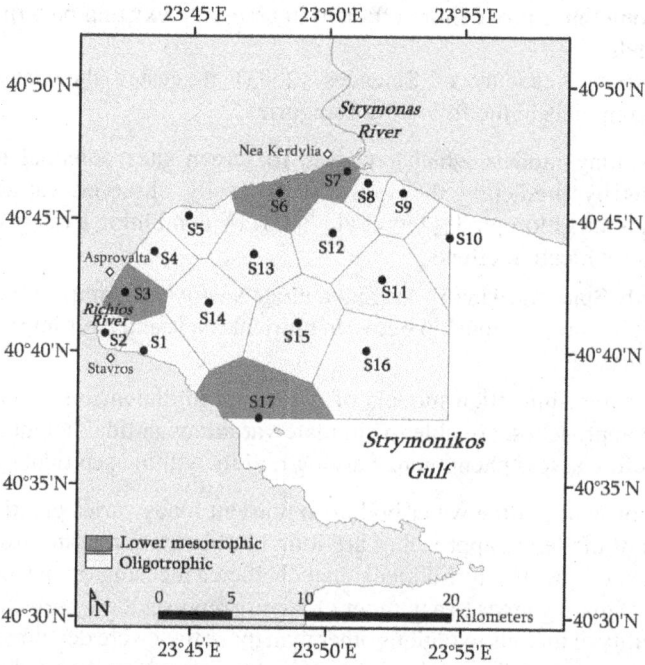

Figure 2.4 Thematic map of eutrophication levels in Strymonikos Gulf by application of the Regime MCA method by assigning weights to the criteria (chla > N-NO$_3$ > P-PO$_4$).

related parameters are useful. The difference of the ranking between cases (i) and (ii) reveals the sensitivity of the method to the assignment of weights to the criteria. Therefore, the procedure of weighting should be carefully applied and the degree of subjectivity minimized. Thematic maps of marine eutrophication, as those produced here, are useful to coastal managers and highly support the decision-making process.

2.10 Modeling in Marine Eutrophication

In marine eutrophication studies, models are used to explore the relationship between causes and effects, to simulate the physical, biogeochemical and biological processes and their interactions, to forecast existing trends in a coastal area as well as evaluate management alternatives and support decision-making and planning.

Two basic types of mathematical models have been developed (Rast et al., 1983); 'Dynamic' and 'Statistical' models. Dynamic models describe the interactions among biological, chemical and physical processes related to aquatic plant growth. They are not widely applicable to any water body, since they are calibrated only for a given water body. Statistical or empirical eutrophication models quantify cause-effect relationship; however, without considering every component involved in the eutrophication process. A variety of such models have been developed (Vollenweider, 1976; Jones and Lee, 1982). Coffin et al. (2018) examined dissolved oxygen as an indicator of eutrophication status in shallow estuaries and stated that simple empirical models can sometimes prove more efficient in assessing existing patterns than more complex models.

The National Academy of Sciences (2003) suggested the classification of eutrophication models in the following categories:

(i) The Screening models which are used to screen sites potential to eutrophic conditions by predicting the value of an easily observed variable such as phytoplankton chlorophyll (Tett et al., 2003) by combining a range of factors to assess eutrophication effects.

(ii) The Steady State and Tidally Averaged models which are simple models capable of describing the relationship between nutrient loads and their impact on marine ecosystems.

(iii) The Dynamic Simulation models of one or more dimensions which are more complex approaches, are able to simulate variations in tidal height and velocity and therefore assess phenomena varying rapidly within each tidal cycle.

The response of marine water bodies to nutrient loads varies greatly; therefore, a wide range of modeling approaches are found in literature ranging from simple to more complex, of one or multi-dimensional character focusing on the assessment of the response of marine areas to a number of factors represented by selected variables. The applicability of models to regions other than those they were developed for, is also an important issue, since the mechanisms related to eutrophication can be affected by the geomorphological characteristics of marine areas (Becher et al., 2000). Scheffer et al. (2003) developed a model with two ordinary differential equations for assessing regime shifts in shallow aquatic ecosystems. Zaldívar et al. (2009) proposed a modeling approach to analyze regime shifts of primary production in shallow coastal ecosystems and to identify possible nutrient thresholds causing shifts between alternative stable states. Cosme et al. (2014) proposed a global method including a biological model of carbon export, consumption, and respiration, to quantify the response of marine coastal waters to nitrogen inputs. Desmit et al. (2018) proposed modeling of the land-ocean continuum to quantify the impact of changes in land use on marine eutrophication. An extended review on modeling marine eutrophication is presented by Ménesguen and Lacroix (2018) with an analysis of 291 representative references and a description of the evolution of tools, their strengths and weaknesses. In addition, Morelli et al. (2018) have published a critical review of eutrophication models for Life Cycle Assessment, where a number of models are analyzed and their advantages and limitations presented.

In 2009, Nixon (2009) argued that eutrophication should be studied at a scale larger than the individual ecosystem for including the majority of the forces that lead to this phenomenon. In addition, in 2001, Cloern stated that eutrophication needs sustained programs of integrated research and monitoring, to prevent and effectively face the damage of coastal regions. It is stated (Diaz and Rosenberg, 2008) that excess of nitrogen during the anthropogenic fertilization of marine systems can lead to an ecosystem response similar to coastal upwelling. Therefore, research efforts should also incorporate the social and anthropogenic pressures beyond the understanding of the environmental and ecosystem processes. The latter is of major importance since decision-making to protect marine areas is based not only on the scientists' proposal but also on public opinion.

Nunneri and Hofmann (2005) proposed a participatory approach based on the DPSIR (Driver-Pressure-State-Impact-Response) analytical tool to evaluate a variety of measures to reduce the input of nutrients and prevent marine eutrophication. DPSIR uses a core set of indicators for environmental performance, including eutrophication, and human activities are considered as an integral part of the ecosystem (EEA, 2001; Zaldívar et al., 2008).

In this framework, a number of systems of global character have been developed such as LOICZ, GOOS, GLOBEC and IMBER. The Land-Ocean Interactions in the Coastal Zone (LOICZ) approach is a core project of the International Geosphere-Biosphere Programme (IGBP) and the International Human Dimensions Programme on Global Environmental Change (IHDP) that deals with global environmental change. LOICZ involves scientists who develop protocols and tools that support the assessment of site-specific and global coastal processes by investigating changes in the biology, chemistry and physics of coastal zone (Crossland et al., 2005).

The Global Ocean Observing System (GOOS) is a global system for observations, modelling and analysis of marine and ocean variables to support forecasts based on marine environmental conditions and social impacts. The Chlorophyll Global Integrated Network (ChloroGIN) sponsored by GOOS, promotes the combination of *in situ* measurements of chlorophyll with satellite data. GLOBEC, a study of Global Ocean Ecosystem Dynamics was initiated in 1990 including a number of Regional Programmes and National Activities *"to advance our understanding of the structure and functioning of the global ocean ecosystem, its major subsystems, and its response to physical forcing so that a capability can be developed to forecast the responses of the marine ecosystem to global change"*. The Integrated Marine Biochemistry and Ecosystem Research (IMBER) project focused on a comprehensive understanding and accurate prediction of ocean responses to global change and the consequent impact on the Earth System and human society.

It should be mentioned that since models are simplifications of reality, it is quite difficult to incorporate all parameters related to the eutrophication process; therefore, most of the times models focus on partial aspects of the phenomenon. In addition, the applicability of a model to various areas is not evident without re-calibration and collection of additional data related to the particular characteristics of the areas. Finally, models can be developed and incorporated in a GIS environment and therefore be combined with spatial analysis methods and mapping techniques in order to produce tools of high integrity for assessing marine eutrophication. Such an approach has been implemented by Lu et al. (2014) who developed a GIS-based system (MWQ-FES) for marine water quality assessment including a marine eutrophication assessment module.

2.11 Marine Eutrophication Monitoring

Although marine water quality monitoring is a complex approach and there is not a specific methodology, it is useful to present the general principles, objectives and methods in use. Nevertheless, data processing applied for the analysis of data from monitoring programs are the methods that have already been described in the current chapter. An early definition of monitoring has been given about half a century ago (MIT, 1997): *"monitoring is defined as the systematic observation of parameters related to a specific problem designed to provide information on the characteristics of the problem that changes with time"*. The set of data collected through a monitoring program can support a number of data processing practices: (a) it can support conceptual and numerical simulation modeling (future prospects) (b) time series analysis (trend monitoring) (c) statistical comparisons (impact assessment) and (d) synthesis and interpretation of the information (NRC, 1990). There are two main monitoring strategies: (a) the "compliance monitoring" usually connected with the implementation phase of a law (national legislation) or an international convention (Frank, 2007) and (b) the "trend monitoring"; the main objective of this trend monitoring is to evaluate the effectiveness of a regulatory framework and produce "benchmarks and performance" (Bell et al., 2008). A monitoring program, although is a purely scientific piece of work including planning of the monitoring program, designing in detail the field work, the analytical laboratory procedures and the data processing, the end users are the policy makers for deciding and implementing coastal management practices (Karydis, 2015). This is why scientists are becoming "participants" in policy making "interfacing" between science and policy. According to Brown (1992): *"the problem then is to visualize and create linkages between the search for the scientific truth and the desire to achieve justice in our society.... The scientific community must seek to establish a new contract with policy makers, based on the demand for autonomy and ever-increasing budgets, but on the implementation of an explicit research agenda.... Scientists and policy makers must work together to make certain that research programs stay focused on policy goals"*.

It is necessary that decision makers and politicians should have a scientific background so that they can communicate with scientists and be able to understand marine water quality issues. Communication between the two sides means therefore that scientists will be politicized to an extent and at the same time policy makers will become technocrats in environmental issues. Scientific and political aspects of marine water quality monitoring cover a wide field of scientific expertise including social and political components which are beyond the objectives of this book. There are a number of review articles for further reading in marine environmental monitoring (Segar and Stamman, 1986; Duarte, 2009; Douvere and Ehler, 2011; Karydis and Kitsiou, 2013; Karydis, 2015). However, some monitoring aspects that will be mentioned in the following chapters as "marine eutrophication monitoring" in the regional seas form indispensable parts of international conventions for the protection of the marine environment. Implementation of measures by these conventions require monitoring schemes to assess efficacy and effectiveness of the measures.

A crucial step in marine monitoring is the planning of the work. As these projects are planned to run for decades, any mistake or inappropriate choice in both field and laboratory work can be very costly not only from the financial point of view but also from the point of the scientific validity of the work. Details on designing protocols for the field work and any possible shortcomings have been presented in a review article by Karydis and Kitsiou (2013).

2.12 Aquatic Microcosm Systems in Marine Eutrophication Studies

The information collected during field work in the marine environment is valuable because it refers to the real system. However, it is difficult to understand complex ecosystem processes in the marine environment and the way system variables interact with each other. In addition, it is not possible to repeat an experiment in nature as environmental conditions are changing constantly. A useful tool for understanding ecosystem processes and interpreting field data in a better way is the use of microcosms. The advantage of using microcosms is that researchers can design experiments, repeat them and work with multispecies cultures simulating phytoplankton communities. The most widely accepted definition of microcosms has been given by NAS (1981). Microcosms can be defined *"as samples from natural ecosystems housed in artificial containers and kept in laboratory environment. These systems are generally initiated by taking whole samples from ecosystems into the laboratory"*. A microcosm system according to UNESCO should fulfill the following three criteria (Anonymous, 1991): (a) it has to be physically confined (b) it is preferable to be of multitrophic character and (c) the volume of the microcosm system should be sufficient so as to allow meaningful sampling; the usual volume of laboratory microcosms varies from 10–20 l up to 1 m³. Microcosm systems have had a significant contribution to understand phytoplankton dynamics, competition experiments, testing aspects of community theory and modelling natural processes (Karydis, 2015). Methodological aspects are given in Table 2.5. Although it seems to be a relatively simple approach, both the design of a microcosm system and the running of the experiment should be carried out with caution Alcazar et al. (1989). Microcosms should not be considered as small scale reproductions of the natural environment but only as a tool to understand physical, chemical and biological aspects related to eutrophication. Spatial scaling (volume of microcosm) and time scaling (duration of the experiment) are the main reasons that make replication of the natural conditions impossible. During planning, the researcher should take into account and make decisions on: boundary conditions, light, temperature, depth, ecological complexity and conditions referring to system's initiation. A microcosm system is a system less complex than the natural environment but more complex than a unialgal or bialgal laboratory culture. The idea for researchers is to simplify the natural system mainly by reducing noise; at the same time to increase the complexity of a microcosm by carrying out multispecies experiments until a compatibility between the field information and the understanding of ecological processes can be achieved. A flow chart (Figure 2.5) explains the evolvement of ecological hypotheses, testing by combining field information with understanding marine processes in multispecies experiments.

The microcosm approach can respond to a number of issues related to marine eutrophication: species competition (Sommer, 1989) and species succession (Sommer, 1988), algal blooms, HABs' dynamics, biogeochemical cycles, sediment and water column interactions, inputs of nutrient loads, system shifts (Liu et al., 2012) and species responses under eutrophic conditions (Zhang et al., 2008; Tsirtsis and Karydis, 1996). It is obvious that microcosms are powerful tools in understanding aspects of marine eutrophication (Table 2.5) but detailed description is beyond the

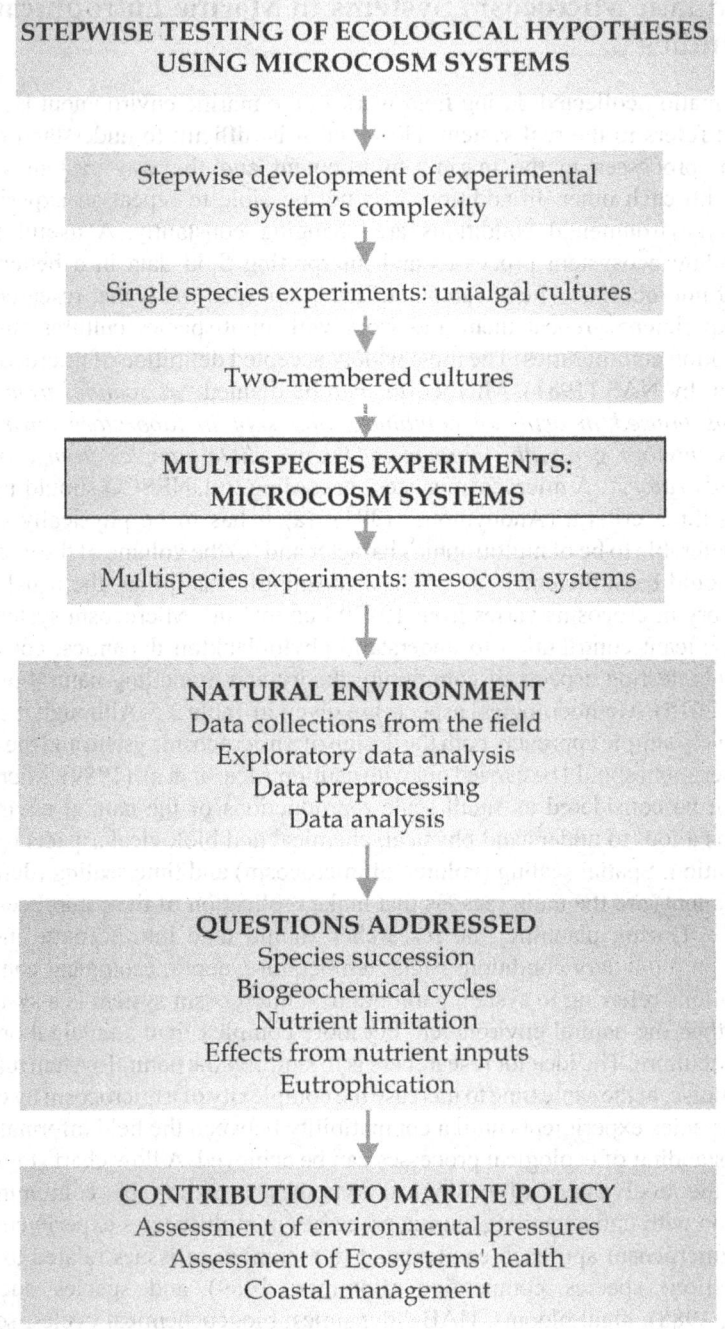

Figure 2.5 Flow chart indicating the role of microcosm systems in environmental assessment and ecosystem' health studies as well as some fields of application and their contribution to marine policy.

Table 2.5 Applications of microcosms in understanding aspects of eutrophication.

Processes related to eutrophication	References
Nutrient loading	Estrada et al. (1996) Petersen et al. (1997)
Physical factors (light, temperature and transparency)	Petersen et al. (1997)
Turbulent conditions	Petersen et al. (1989)
Effects of "spiking" (nutrient additions during the period of the experiment	Buyakates and Roelke (2005)
Community response to nutrients	Tsirtsis and Karydis (1997)
Community component interactions	Vaque et al. (1989)
Ecological Processes in phytoplankton ecology (competition, succession)	Sommer (1988, 1989, 1991) Liu et al. (2007)
Modeling trophic conditions	Roelke et al. (2003) Tsirtsis and Karydis (1997) Bartleson et al. (2005) Domis et al. (2007)

scope of this book. Readers can find more information on textbooks: Beyers and Odum, 1993; Wiens, 2001; Estrada and Peters, 2002; Karydis, 2014. There are also many papers referring to nutrient limitation (Table 2.5), microcosm bioassays at different N/P ratios to evaluate nutrient limitation (Buyakates and Roelke, 2005; Estrada et al., 1996; Estrada and Peters, 2002), the role of sediment and experiments at microcosm level used for modeling eutrophication processes.

3

Eutrophication and Governance of the Regional Seas

3.1 General Background

As there are no physical boundaries in the marine environmental pollution problems that include loss of diversity, toxic and hazardous wastes, sewage related issues, depletion of biological resources and pollution from industrial as well as agricultural sources, legal regulations at an international level have been taken (Sands et al., 2012). The difficulties for the various states to come to an agreement, at least at a scale of regional seas is threefold: (a) the marine environment is particularly complex where many factors and processes are actually interconnected. An international legal scheme that will allow effective management requires that these related factors be studied and their possible interactions should be understood; this means that these factors should not be treated as discrete cases (b) these problems cannot be addressed properly unless economic, political, social and cultural factors have been taken into account and (c) the understanding of a number of non-legal factors: these are the scientific consensus regarding causes, effects and seriousness of the problem; the existence of previous relevant multistate agreements; possible short term political benefits and partners' perception that they are addressing the problem by doing their "fair share". In addition, possible economic costs during the implementation phase and the number of states involved in a particular agreement are factors decreasing the degree of likelihood for reaching an agreement.

As there is usually a cost when environmental regulations are set, the community should be convinced by the scientific estate that the proposed measures are based on a sound scientific basis, necessary to prevent environmental impacts. In addition to the scientific background of the agreement, the partnership between environmental protection and economic development should be taken into account. Although the implementation of environmental laws on a short-term basis usually imposes considerable economic costs, in the long run, many benefits concerning the economy may emerge, especially among countries that are developed from the technological point of view or the quality of their environment as an asset can be a significant resource; the latter is the case with tourism and biological marine resources.

During the last twenty years, in addition to science and economics, a third player has emerged: it is the participation of the public as an influential factor in environmental decision making as well as in setting environmental disputes. Public participation materializes through individuals and indigenous communities, the media of communication (press, television and internet), associations of lawyers, groups of interest like chamber of commerce and private groups. More information about the legal status, rights and functions of this type of groups are given in Section 3.2.

3.2 Governance

The results of activities, initiatives, conflicts or even overlapping between the institutions mentioned above, is an aggregation of laws, policies, rules and norms aiming at a common objective which is the protection and management of the marine environment. As a collective term to encompass all types of regimes and actions, the term "cluster" has been coined. In particular, DiMento and Hickman (2012) have designated the term cluster to describe the aggregation of attempts to improve the quality of regional seas. The totality of activities deriving from these legal tools, seeking to provide "effective management of the seas is known as governance". To understand how the complex system of legal and administrative environmental tools implemented in the marine environment works, the actors participating in environmental issues and their potential and influence should be presented shortly: these are international organizations, coastal states and non-state actors.

States are the primary components for any international legal framework as they contribute to the creation, implementation and evaluation of the effectiveness of legal measures. They are also members of international organizations and they support the participation of various non-state actors in the formulation of rules and the decision making. There are currently almost 200 member states within the UN. Based on economic indices, competent organizations have divided the states into developed countries as well as economies of transition. This heterogeneous blend of states characterized by different economic, social, political, geographical and ecological conditions results into conflicting interests among these states. It is therefore obvious that coming to an agreement for marine environmental protection and management is both a "delicate and painful procedure".

The second component of governance is international organizations aiming at marine environmental protection. They are established at various levels namely global, regional, sub-regional and in some cases bilateral levels. The majority of the international organizations that will be mentioned in the current chapter have responsibility for the application or even enforcement of international obligations; they have also competence for the development, assessment and amendment of their regional agreements. They are categorized into global organizations connected with the UN, organizations outside the UN system and organizations functioning under other treaties (Sands et al., 2012).

The functions of international organizations are related to the development, management and generally administrative issues of the marine environment. There are four main functions regarding the international organizations: (a) they provide to the states a platform that can be used for communication among member states, consultation, sharing of ideas and consensus, wherever possible, at regional or even international level. In addition, they contribute to the formation of an international agenda regarding the marine environment and the encouragement of environmental marine research (b) the second function is the collection, sharing and dissemination of information referring to marine environmental quality, practices and measures. They can also act as a "warning system" for emergencies such as toxic algal blooms and various sea accidents that can have effects on the marine environment (c) international organizations can also agree on policy initiatives and standards, they may adopt rules accepting obligations or establish new institutional arrangements and (d) another function is the setting up of a mechanism to settle disputes among states.

International organizations have also been active in setting international standards and developing treaty language; in addition, they have been acting as *"watch dogs"* during the implementation phase of treaty commitments. The non-state factors have been influential players in international environmental issues. Five categories of non-state factors seem to play a dominant role in regional and marine environmental affairs (a) non-profit making associations in environmental affairs usually NGOs (b) private companies (c) legal organizations (d) academic communities and (e) individuals. Participation of non-state factors has also been encouraged by Agenda 21 (1992) by asking member states to intensify their contribution in *"policy design, decision making, implementation and evaluation at the individual agency level, in inter-agency discussions and in United States conferences"*. It has been possible for non-state actors to participate as observers in international organizations. It has to be mentioned that the OSPAR Convention (1992) for the protection of the North Sea and NA Atlantic, for the first time included a treaty provision for observers; this Convention does make a distinction between them and the States, international organizations and non-governmental organizations regarding their rights to participate in activities that are related to the Convention.

As any Convention regarding the marine environment is a compromise initiated by political, social, economic interests and scientific evidence, it is obvious that science is the only objective element throughout the whole procedure. This makes the role of science imperative during the phases of policy design, decision making and implementation. Scientific information can be generated from various sources including national research centers, government departments and non-state sources as well as from international processes such as GESAMP. The significant contribution of the environmental estate has a two-way effect pointed out as early as 1990: *"environmental movement has been powerfully affected by the consequences of science misused to the detriment of the living world, but even more importantly by what advancing science has revealed about the structure and process of nature"* (Caldwell, 1996). On the other hand, there are many analysts who maintain that the impact of science on environmental policies is limited. It has been supported

(Engelhardt and Caplan, 1987) that controversies regarding environmental issues are finally ending as scientific controversies. This is because in many cases the "non-scientific factors such as cultural, social, ethical or economic factors are the most important and dominating over environmental issues". These are called *"scientific controversies with an overlay"*. Apart from the non-scientific factors mentioned above, there are fairly often discrepancies within the scientific community itself as well as difficulties between the scientific and the political estate. Matching science with policy has always been the weak interlink in environmental policy issues (Karydis, 2015).

The private sector such as companies and associations have set up proposals for the development of marine environmental conventions and at the same time they hold *"regular dialogues"* with international organizations. Their contribution in marine environmental issues is threefold (Sands et al., 2012): (a) to support a precautionary approach to marine environmental challenges (b) to undertake initiatives relevant to environmental responsibility and (c) to encourage the development and use of technologies friendly to the marine environment. According to the OECD (1984) guidelines, as updated in 2000 (DAFFE/IME, 2000): *"enterprises should within the framework of laws, regulations and administrative practices in the countries in which they operate, and in consideration of relevant international agreements, principles, objectives, and standards, take due account of the need to protect the environment, public health and safety, and generally conduct their activities in a manner contributing to the wider goal of sustainable development"*. The extending relationship between human rights and environmental issues at international level, has encouraged the citizens to acquire the right to complain to international bodies. The role of citizens of nation-states is enhanced as in many cases they can report violation of the agreements by governments to international organizations and to national competent authorities. Citizen's rights are reflected in the Rio Declaration which recognizes the right of the citizens' participation in decision-making processes as well as to be able to have access to relevant information. According to Principle 10 of the Rio Declaration: *"environmental issues are best handled with the participation of all concerned citizens, at the relevant level. At the national level, each individual shall have appropriate access to information concerning the environment that is held by public authorities, . . . and the opportunity to participate in decision-making processes"*.

Last but not least among the non-state actors is the contribution of the media. The media can focus on specific environmental issues that have an influence on the public opinion making the states change their position. They also play a leading role in the understanding of the problem and possible actions as well as the mobilization of the public opinion.

The cluster of actors involved in policy making and implementation as well as many different types of marine pollution, indicate that marine eutrophication cannot be studied efficiently as an isolated problem but in connection with the other pressures and causes within the complex legal framework set by international, regional, national or even bilateral agreements for marine environmental protection and management.

3.3 Marine Eutrophication Policy

The role of science in understanding, assessing and proposing the necessary steps in policy measures has already been mentioned in the previous sections. Science is also contributing expertise in the so-called policy life cycle consisting of three phases (de Jong, 2006): (a) the phase of the discovery of the problem. This phase is characterized by a high degree of uncertainty, sometimes defiance regarding the extent of the problem or even denial about the problem itself. It is obvious that at this phase the problem of eutrophication in an area cannot be accepted as a political issue. (b) Marine eutrophication is finally accepted as a pollution problem and the issue is then placed in the agenda. This is the political phase of decision-making. Different actors negotiate for the possible measures and finally political decisions are made and (c) this mature stage is characterized by environmental management and implementation of measures. These include normative issues, especially on land-based activities, marine environmental monitoring and if necessary amendment of the measures and more legal measures. At this phase, the tools of science decrease the uncertainty about the problem's parameters and sets the basis for a better control over the particular issue.

The key factor regarding the influence of science in policy making is consensus within the science estate. This rarely happens as "science itself, because of its own internal propensity for conflict, can only make limited contribution to the resolution of international conflicts over public choices" (Boehmer-Christiansen, 1989). Although modest critique is advancing scientific knowledge, heavy critique is counterproductive and actually non-necessary as what is known as "error-costs" low (Collingridge and Reeve, 1986).

The uncertainty in environmental policy stems from two factors: the ecosystem's complexity and the enormous volume of environmental data that have to be processed and assessed. The former is connected with the fact that scientific knowledge used for public policy known as regulatory science (Jasanoff, 1990) often works "*at the margins of existing knowledge, whereas academic science works within established paradigms*". The indecision in policy making due to uncertainty is getting a major problem for large-scale environmental issues at regional or international level. The complexity of the ecosystems has not allowed so far the structuring of a general ecosystem theory and it is obvious that this is also reflected in the marine environment. For the average ecologist, the term ecological theory refers to a set of various ideas and constructs (Pianka, 1981). Ecological theories have been classified either as explanatory models or as tools for quantitative approaches (Loehle, 1983). To the uncertainty due to weakness of the ecological theory used for environmental quality assessment and policy making, a number of methodological errors has to be added to the system: poor experimental design and sampling bias (Karydis and Kitsiou, 2013), experimental artifacts and technical malfunction are among the main causes in the practical side of the use of ecology in policy issues (MacArthur, 1972). The key issues in environmental marine policy are the hypotheses regarding ecosystem's function, ecological theory and as the last step, the incorporation of the ecological conclusions into the legal tools for the protection and management of the marine environment.

The role of science is critical at the different phases of the eutrophication phenomenon including recognition of eutrophication as a political problem, policy making and management (Winselmius, 1986). The first step is the discovery phase. Eutrophication issues are accepted as a problem aiming at the politicization of the issue. There is usually a lot of controversy as far as the importance of the problem is concerned. The scientific background already set up leads to the second step which is the setting of the problem on its political dimensions. The problem "matures" in the form of a legislative framework. The policy life cycle is completed by the implementation phase. However, things are not usually as linear as presented above. Marine eutrophication as an environmental problem changes in its structure with time not necessarily because of the uncertainties already mentioned but also because of changes in social attitudes, technological developments and advances in science. Jasanoff as early as 1990 had noticed that the scientific contribution in policy making is not a "one-shot process". This means that during the implementation phase, many rounds of consultation take place as a reaction to new scientific knowledge, societal perception and feedback regarding fiscal aspects. It is a common practice for many regional conventions to be revised from time to time, incorporating the amendments on subjects gained by experience. The key issues in environmental marine policy are the hypotheses regarding ecosystem's function, ecological theory and as the last step, the incorporation of the ecological conclusions into the legal tools for the protection and management of the marine environment.

According to de Jong (2006), the role of science in decision making focuses on three issues: the normative, the structural and the temporal issues. The normative aspect maintains that science is imperative for decision making and management, in spite of any uncertainties due to the highly complicated nature of marine eutrophication due to: (a) the openness of the system (b) the dynamics of physical, chemical and biological processes and (c) the fact that both nutrients and chlorophylls are built-in variables in the natural system and therefore it is difficult to discriminate between natural causes and human-induced nutrient loading (Kitsiou and Karydis, 2011). Generally speaking, the policies to eliminate marine eutrophication are focusing on the reduction of nutrient inputs from terrestrial sources, construction of sewage treatment facilities and encouragement of the farmers to apply good agricultural practices. The structural issue refers to the interfacing between science and policy (Karydis, 2015). Matching science with policy has not been an easy target, although during the last decade communication between the two sides has improved and seems that they understand each other better. This is possibly due to the fact that scientists are more and more getting "politicized" and policy makers are developing technocratic attitudes (Karydis, 2015). Lastly, the temporal aspect refers to the policy life-cycle, in other words how science is influencing matters once it has been incorporated in the policy process.

The need for taking measures to protect the marine environment became obvious since the sixties when a number of sources that cause deterioration of the marine environment had been identified. Among these sources, land-based pollution, untreated sewage, eutrophication, invasive species, habitat destruction and hazardous substances were considered as the most important. The first rules referring to the protection of the sea from pollution were developed at a global and regional level as early as 1972 (Sands et al., 2012). However, the United Nations Convention on the Law of the Sea (UNCLOS) signed in 1982 aimed to establish *"a legal order for the seas and oceans which will facilitate international communication, and will promote the peaceful uses of the seas and oceans, the equitable and efficient utilization of their resources, the conservation of their living resources, and the study, protection and preservation of the marine environment"* (Sands et al., 2012). The UNCLOS, entered into force in 1994, requires that state parties should take measures to prevent, reduce and control marine pollution. Rules on information, monitoring, scientific research and water quality have been included. The protection and preservation of the marine environment is addressed specifically in the 3rd Section of the UNCLOS. Article 194(3) mentions explicitly the obligation of the member states to prevent pollution, specifying sources such as land-based activities, seabed activities, dumping and pollution from the atmosphere. It is noteworthy that all these sources are closely connected with marine eutrophication (UNEP, 2003a; Karydis and Kitsiou, 2012). Article 194(5) refers to the protection of rare and fragile ecosystems, the protection of habitats and the endangered species. All these problems can also be caused by marine eutrophication.

The measures mentioned above for marine water quality and ecosystem's protection together with detailed standards formed a spectrum of rules for developing and implementing scientific research, monitoring programs and marine environmental assessments (Karydis, 2017). It is also necessary to provide technical assistance wherever needed. However, UNCLOS is only a general framework setting a comprehensive legal order for the regional seas and oceans. The physical characteristics of every marine environment including regional seas, the level of development of the surrounding countries and the environmental problems that have meanwhile been accumulated, has made governance a particularly complex issue that can function efficiently only at a regional level.

3.4 Regional Arrangements

The main "platform" for marine environmental protection and assessment at a global scale is the UNEP's Regional Seas Programme; UNEP encompasses thirteen regional seas programmes and the Framework Conventions for the Black Sea, The Northeast Atlantic, the Caspian Sea and the Polar Regions. A total of 149 states, representing 95 per cent of the world's states, are part of the programme, covering 18 regions of the World (DiMento and Hickman, 2012). These marine areas include the Mediterranean Region, the Gulf Sea Area (ROPME), the Western Africa Region, the South-East Pacific Region, the Red Sea and Gulf of Aden, the Wider Caribbean Region, the Eastern Africa Region, the Pacific Region, the Black Sea Region, the North East Pacific Region and the East Asian Seas (COBSEA). UNEP was established after the Stockholm United Nations Conference on the Human Environment in 1972. UNEP is the main institution of the United Nations in environmental affairs. It has been described as the *"voice for the environment within the United Nations system"* and according to the United States Secretary General as the *"environmental conscience of the United Nations"* (Desai, 2006). The Regional Seas Programme is supported by fourteen organizations of the United Nations, thirty non UN regional organizations and more than a thousand national institutions. Among the draft principles adopted by UNEP since 1978 (Sands et al., 2012), Principles 3 and 4 require from the member states to "make environmental assessments" whereas Principles 7 and 8 include requirements on scientific studies.

The main objectives set by the Governing Council of UNEP in the Regional Seas Programme were the development of extensive action plans for the protection and management of regional seas areas. The proposed strategies included (Kaniaru, 2000): *"Promotion of International and Regional Conventions, guidelines and actions for the control of marine pollution and for the protection of aquatic resources; assessment of state of marine pollution of the sources and trends of this pollution and of the impact of pollution on human health, marine ecosystem and amenities; coordination of the efforts with regard to the environmental resources; support for education and training efforts to make possible the full participation of marine and coastal resources"*. A summary of the main elements of these regional programmes is given in Table 3.1. Regional Seas Programmes differ in a number of ways: the length of the coastline, the jurisdictional zone, biological resources, mineral resources, coastal resources and the level of development of the contracting states.

3.5 International Conventions and Marine Eutrophication

3.5.1 *The Mediterranean Sea*

The Barcelona Convention is the oldest UNEP agreement for regional seas. It was adopted in 1976 and entered into force in 1978. The objective of the Convention is to "*reduce pollution in the Mediterranean Sea, protect and improve the marine environment in the area, thereby contributing to its sustainable development*" (Birnie et al., 2009). The original Convention has been updated by amendments following the experience gained by the years and the effectiveness of the mandatory measures included in its Appendices. The main characteristic of the Mediterranean Sea is the high social, economic and political diversification among 22 signatory Member States. There are also many legal complexities regarding the extent of their territorial waters, the boundaries of the fishing zones, national legislation for zones of ecological protection and adoption of rights for the historical and archeological objects within the 24-mile limit from the baselines of the territorial waters. The Mediterranean Action Plan (MAP) has formed the framework for protocols and annexes regarding the protection of the Mediterranean System; this cluster of agreements, laws, rules and limitations is known as the "Barcelona System".

Following the Rio Conference (1992) on Environment and Development, the Barcelona Convention underwent important changes. The "*Mediterranean Action Plan*" was renamed as "*Action Plan for the Protection of the Marine Environment and the Sustainable development of the Coastal Areas of the Mediterranean*" (MAP Phase II). Some of the existing protocols were amended following new concepts about the marine environment whereas other covering new subjects were introduced. There are seven protocols under the Framework Convention. These include the "*Protocol Against Pollution*", the protocol for "*Prevention of Pollution in the Mediterranean*

Table 3.1 Conventions of Regional Seas and their relevance to marine eutrophication.

Region	Convention	Action plan	Relevance to eutrophication
Mediterranean	Barcelona Convention 1978/2004	Mediterranean Action Plan	Eutrophication Monitoring and Assessment is materialized through the MED POL Programme as marine eutrophication is a component of MED POL.
Black Sea	Bucharest Convention 1992	The revised strategic Action Plan for the Environmental Protection and Rehabilitation of the Black Sea 2009	There are two protocols of the Bucharest Convention relevant to marine eutrophication: Protocol I referring to Land Based sources and Protocol IV on Biodiversity and Landscape.
Baltic Sea	Helsinki Convention 1974/1992	Baltic Sea Comprehensive Environmental Action Programme 1992	The Action Plan adopted by the ministers of the Baltic countries requires, inter alia, specific measures for eutrophication: (a) waste water treatment for phosphorus removal up to 80-90% (b) reduction of nitrogen inputs of agricultural origin and (c) nutrient reduction from the River Basins.
Caspian Sea	Tehran Convention	Caspian Environment Programme 1992	Eutrophication is not mentioned in the Convention
North Sea and NA Atlantic	OSPAR Convention 1998	OSPAR Action Plan 1998–2003	The second priority of the OSPAR Convention refers to nutrient reduction "where the inputs are likely to cause eutrophication". ASMO's Report proposed common procedures for marine eutrophication: "Common procedure for the identification of the Eutrophication Status of the Maritime Area of the Oslo and Paris Conventions".
East Asian Seas	No Convention There is a Coordinating Body of Seas of East Asia (COBSEA)	Action Plan for the Protection and Development of the Marine and Coastal Areas of the East Asian Region 1981/1991	The Programme for Action is also aiming at the protection of the Marine Environment from Land Based activities, including nutrient discharges.
West and Central Africa (WACAF)	Abidjan Convention	West and Central Africa Action Plans	
Wider Caribbean	Cartagena Convention 1986	Caribbean Action Plan	Some aspects of the Convention are related to eutrophication

Table 3.1 contd. ...

...Table 3.1 contd.

Region	Convention	Action plan	Relevance to eutrophication
The Gulf (ROPME Sea Area)	Kuwait Convention 1979	Action Plan for the Protection of the marine environment and coastal areas of ROPME	
Red Sea and Gulf of Aden	Jeddah Convention 1985	Action Plan for the Red Sea and the Gulf of Aden 1982/1995/2005	Threats concerning eutrophication are mentioned indirectly through the land-based sources of pollution, dredging and airborne pollution
South East Pacific	Lima Convention	South East Pacific Action Plan	
South Pacific	Noumea Convention	No Action Plan	
Indian Ocean (Eastern Africa)	Nairobi Convention	East Africa Action Plan	

by *Dumping from Ships and Aircrafts*", the protocol on "*Combatting Pollution from Oil and Other Harmful Substances*", the protocol for the protection against pollution from "*Land-based Sources*", the Protocol for the "*Protection of Specially Protected Areas*", the protocol on "*Pollution Resulting from Exploration and Exploitation of the Continental Shelf*", the protocol for the protection from "*Transboundary Pollution*" and the "*Integrated Zone Management*" Protocol. All these protocols are directly or indirectly related to marine eutrophication.

From the point of view of eutrophication, the Dumping Protocol prohibits any disposal, with exception of wastes, due to routine operations of vessels. There are also some exceptions referring to fish waste, dredged materials in inert uncontaminated geological materials. These categories covered by exceptions, also contribute to eutrophication. The Land-based Protocol refers to effluents deriving from terrestrial activities. They can be either point pollution or diffused sources. The discharges come from run-off of the land, river flow or sewage–industrial outfalls. The protocol lists 19 categories of pollutants, some of them connected with eutrophication.

3.5.2 The Black Sea

The Black Sea has faced serious pollution problems including eutrophication. In 1992 a Convention for the protection of the Black Sea was signed among six states namely Bulgaria, Georgia, Romania, the Russian Federation, Turkey and Ukraine in Bucharest therefore known as the Bucharest Convention. The main objective of the Convention is "to substantiate the general obligation of the Contracting Parties to prevent, reduce and control the pollution in the Black Sea in order to protect and preserve the marine environment and to provide legal framework for cooperation and concerted actions to fulfill this obligation". There are three protocols accompanying the Convention: (a) the protocol referring to land based sources of pollution (b) the dumping waste protocol and (c) the protocol regarding joint action of accidents. Protocols (a) and (b) are directly related with eutrophication.

3.5.3 *The Baltic Sea*

The limited communication between the Baltic Sea and the North Sea through the Skagerrak and Kattegat Straits, renders the marine area sensitive to pollution and particularly to eutrophication which is ranked as the worst threat to the Baltic. The first regional seas treaty covering all sources of marine pollution was the Helsinki Convention for the Protection of the Marine Environment in the Baltic Sea. It was signed in 1992 by all coastal countries. Unfortunately, the first two decades of operation showed that there was no success in mitigating environmental degradation in the Baltic (Fitzmaurice, 1998). The Helsinki Convention was therefore replaced in 1992 by a new agreement with emphasis on the restoration and preservation of the ecosystem integrity in the Baltic Sea. The Convention was beyond pollution assessment to cover aspects referring to good ecosystem health, habitat conservation, biological diversity, ecological processes and the sustainability of the natural resources. Since 2003, the Ministerial Declaration of the Helsinki Convention set priorities for marine eutrophication, diversity conservation and the development of "*ecological quality objectives*". In addition, environmental impact assessments, reporting, exchanging information and providing information to the public, were included in the assessments. Although hazardous substances, marine safety, biodiversity and nature protection are subjects of concern, eutrophication has been declared as an issue of priority (Sands et al., 2012; Hakanson and Bryhn, 2008). From the governance point of view, it has to be stressed that the Baltic Sea contains legal subsystems. All coastal countries apart from the Russian Federation are members of the EU since 2004. This means that a number of EU Directives, dealing directly or indirectly with marine eutrophication are in force by all EU Member States of the Baltic Sea. These Directives and their relevance to marine eutrophication are presented in the Section 3.7 under the title "European Union Policy in Marine Eutrophication".

3.5.4 *The Caspian Sea*

The Tehran Convention for the Protection of the Marine Environment of the Caspian Sea was signed in 2003 by the States surrounding the Caspian Sea: Azerbaijan, Iran, Kazakhstan, the Russian Federation and Turkmenistan (BSC, 2009). It was entered in force in 2006. There are many pollution problems in the area such as nutrient discharges from untreated sewage inflows, oil pollution problems and pollution from hazardous compounds but eutrophication is the principal environmental problem (Karydis, 2017). The main objectives of the Tehran Convention are: (a) reduction and/or prevention of pollution and pollution control (b) protection and remediation of the marine environment (c) measures taken by the contracting parties to restrict over-exploitation of natural marine resources and (d) protection of endemic, endangered species as well as their habitats. Principles embodied in the Tehran Convention are the precautionary principle, assess to information by the public and stakeholders and principle "*the polluted pays*" (Sand et al., 2012).

3.5.5 *The North Sea and NA Atlantic*

The Oslo Dumping Convention (1972), the Paris Convention on Land-Based Pollution together with the Bonn Agreement for *"Co-operation in Dealing with Pollution of the North Sea by Oil"* were replaced by the OSPAR Convention (Oslo + Paris), which is the legal framework for the Protection of the North Sea and the North-East Atlantic (1992). The main objective of the OSPAR Convention is to control and mitigate marine pollution in an integrated and comprehensive manner. The most innovative aspect of the OSPAR Convention was the *"ecosystem oriented approach"*: areas suffering from marine pollution should be restored *"so as to safeguard human health and to conserve marine ecosystems and when possible, restore marine areas which have been adversely affected"*. As the Convention encouraged scientific research, a great deal of scientific information has been produced over the last twenty-five years (Karydis, 2017; Salomons et al., 1988).

The OSPAR Convention is very clear about marine eutrophication: *"The Contracting Parties to the Convention for the Protection of the Marine Environment of the North-East Atlantic adopt the following objective and strategy for the purpose of directing the work of the Commission with regard to eutrophication"*. In addition, Objective 1.1 of the Convention refers exclusively to marine eutrophication: *"In accordance with the general objective OSPAR's objective with regard to eutrophication is to combat eutrophication in the OSPAR maritime area in order of achieve and maintain a healthy marine environment where eutrophication does not occur"*. Eutrophication control had also to comply with the legal principles mentioned above.

3.5.6 *The East Asian Seas*

The East Asian marine environment is classified into three subregions: the Southeast Asia, the North-East Asia and Australia. These subregions are characterized by unique weather patterns, special biological features and economic bases. The East-Asia marine area includes the Philippine Sea, the Sulu Sea, the Timor Sea, the Celebes Sea, The Arafura Sea, the Banda Sea, the Flores Sea, the South China Sea, the Java Sea, the Straits of Singapore, the Straits of Malacca, the Oceans of Australia and the Andaman Sea (DiMento and Hickman, 2012; UNEP, 2008). The physiography of the area is also highly diversified: there are scattered islands throughout and most of the seas are interconnected via channels and straits. There are also deep basins, volcanic islands as well as coral islands (Teng, 2006). In addition to the species diversity, there is also significant ecosystem diversity: seagrass formations, mangrove forests and large coral reefs are the most dominant (UNEP, 2005). The East Asian Seas are characterized by a wide variety of natural resources. However, most of these abundant resources are on a decline. This is mainly due to the economic growth of most of the states in the area. Increase in tourism and industrial activities have affected the marine environment rather seriously.

The Regional Seas Programme for the East Asian Seas, approved in 1981, is the main management tool for the Protection and Development of Marine and Coastal Areas for East Asia. Unlike other Regional Seas Programmes, the East Asian

Programme is not connected with the Convention. However, there is a Coordinating Body of Seas of East Asia known as COBSEA, tasked with the collection of marine environmental data and coordination of national policies regarding the protection of the area. The lack of *"binding rules"* is causing a number of problems in the collaboration among the South East Asian States and therefore the effectiveness of the measures. There are numerous countries refusing to discuss measures or to be tied to any legal agreements. In addition, most of the states are reluctant to contribute financial capital to maintain the relevant functions of the COBSEA (DiMento and Hickman, 2012). Apart from the RSP, there are many bilateral collaborations. An active agency that needs be mentioned is the Partnerships in Environmental Management of Seas of East Asia, known by its acronym, PEMSEA. The emphasis of PEMSEA is placed on coastal zone management throughout the East Asia. Other institutions mainly involved in the South China Sea include IUCN, UNEP and GEF. However, the complexity of the cluster due to many organizations involved, lack of cooperation and coordination, funding constraints and the lack of legal framework allowed very limited success in the improvement of the quality of the marine environment.

Eutrophication does not seem to receive any special attention as such but the land based protocol includes nutrient loading of the marine environment from rivers and non-point sources.

3.5.7 The West and Central African Seas

Twenty-two African States, mainly coastal, make up the West and Central African region (WACAF). The main resources for most of the WACAF nations are fishing, tourism and industry. As most of the economic growth in the area is rapid and undisciplined, it causes impacts on the coastal zone from the overuse of the coastal environment. In addition, transportation of goods by boat inland from the coast contributes to the environmental impacts in the area. The length of the coastline is about 8,000 miles, whereas the population of the region exceeds 300 million people. The overuse of natural resources and the impact on the marine environment have taken the attention of UNEP: *"this has led to the degradation and overexploitation of natural resources, as industry and commerce have generally focused on maximizing profits at the expense of sound environmental management"* (UNEP, 2002).

The WACAF group of states adopted an Action Plan under the aegis of the UNEP Regional Seas Programme. This Action Plan was formulated as a Regional Convention, known as the Abidjan Convention signed in 1981 and set into force in 1984. The signatory countries were Angola, Benin, Cameroon, Cape Verde, Congo, Democratic Republic of Congo, Equatorial Guinea, Gabon, Gambia, Ghana, Guinea, Guinea-Bissau, Liberia, Mauritania, Namibia, Nigeria, Sao Tome and Principe, Senegal, Sierra Leone, South Africa and Togo. The main objectives of WACAF included strict regulation of sewage and solid waste, coastal erosion and habitat protection (WACAF, 2010). All these objectives are closely related to marine eutrophication.

The cluster regarding the governance in the West and Central African marine area is rather complex since various institutions overlap in jurisdiction, causing conflicts among various aspects of marine policy, management and implementation Plans (Kouassi and Biney, 1999). However, it has to be mentioned that the OSPAR Commission has assisted WACAF considerably to develop regional agreements and improve governance. There are many drawbacks regarding environmental protection due to the priority of most of the states in areas of economic growth where environmental priorities are set aside.

3.5.8 *The Wider Caribbean Region*

The Wider Caribbean Region (WCR) is a semi-enclosed water mass connected to the Central Atlantic. By the term "Wider Caribbean Sea", we mean the Caribbean Sea, the Gulf of Mexico as well as a number of bays and minor marine areas (DiMento and Hickman, 2012). The area is characterized by highly diversified ecosystems, including coral reefs with unique coral species and exotic fish species, sea turtles and other interesting forms of marine life. WCR encompasses thirty-three states and territories (UNEP, 2010). However, all these states are characterized by a different level of growth and development. The area covered by the Wider Caribbean Sea is about 15 million square kilometers but only 4.2 million square kilometers is the area of the Caribbean Sea itself (Clarke, 2001). There are three large rivers that is Amazon, Orinoco and Magdalena, discharging directly into the Wider Caribbean Sea. The combination of river water masses and currents bringing water from the Atlantic disperse waste and pollutants over a wider area. It is believed that significant amounts of nutrients are discharged from these rivers into the Caribbean Sea between May and November causing eutrophication (Corredor and Morell, 2001). Phytoplankton blooms have been observed over the last twenty five years causing oxygen depletion (Garzon-Ferreira et al., 2000) and mass fish mortalities (UNEP et al., 2006). Many other places in the Caribbean are suffering from eutrophication; these include among other places the San Huan Bay (Puerto Ricco), Havan Bay (Cuba) and Kingston Harbor (Jamaica) (CEDI, 2000; UNEP et al., 2004). It is obvious that eutrophication is high in the list of priorities for marine environmental protection.

The population in the area is expected to reach 90 million by the year 2020, the population densities being the highest along coastal areas, increasing even more during the tourist season (UNEP, 2006). The main source of income in the Caribbean is tourism as the area has built a high reputation on its unique coral reefs, endemic fish species, clean ocean waters and swimming beaches, attracting about 100 million tourists per year. Intensive touristic, fishing and mining activities in the area have a significant contribution to marine environmental degradation (Colmenares and Escobar, 2002). Tourism is the major threat for the marine environment. The second major pollutant in the area are the nutrients. It has been estimated that at least 13,000 tons of nitrogen and 6,000 tons of phosphorus are released annually in the marine environment (CEP, 2010).

The Caribbean Environmental Programme (CEP) within the Caribbean is the principal legal tool in coastal and ocean protection in the area. It is the core organization for the creation of protocols in the Wider Area. The Caribbean Environmental Programme is the platform for the *"Caribbean Action Plan"* and the *"Convention for the Protection and Development of the Marine Environment in the Wider Caribbean Region"* known as the *"Cartagena Convention"* that came into force in 1986. The Cartagena Convention aims at regulating pollution from ships, dumping, airborne and land-based activities. Among the protocols supplementing the Convention is the *"Protocol of Land-Based Sources and Activities"* (LBS) and the Protocol on *"Specially Protected Areas and Wildlife Convention"*. Both these protocols are related to marine eutrophication. Policy responses on nutrient pollution are mainly focused on the Protocol Concerning Pollution from Land-Based Sources. There are four Annexes in this Protocol: Annex Ic (Associated Pollutants of Concern) is referring to various primary pollutants including nutrients (nitrogen and phosphorus compounds), Annex IIA, is referring to characteristics and composition of the waste as well as emission source control and management factors and Annex III is referring to Domestic Wastewater and Annex IV is referring to agricultural non-point sources of pollution.

3.5.9 The Gulf

The Gulf is protected by the *"Kuwait Regional Convention for the co-operation on the protection of the marine environment from pollution"* that was signed in 1978 by the states surrounding the Gulf (Bahrain, Iran, Iraq, Kuwait, Oman, Qatar, Saudi Arabia and the United Emirates). The main objective of this Convention is to protect the marine environment from *"oil and other harmful or noxious materials arising from human activities on land or at the sea especially through indiscriminate and uncontrolled discharge of these substances"*. However, nutrients are taken indirectly into consideration as Article VI on *"Pollution from Land-Based sources"* is mentioning the need of measures against pollution, resulting from outfalls. Pollution from dredging (suction dredging and coastal dredging) and the appropriate measures to be taken, are included in Article VIII: *"Pollution from other human activities"*. In addition, monitoring programmes should be developed related to all types of pollution (Article X: Scientific and Technological Co-operation).

3.5.10 The Red Sea and the Gulf of Aden

The Regional Convention for the Conservation of the Red Sea and the Gulf of Aden Environment, known as the Jeddah Convention was signed in 1982. The geographical coverage of this Convention includes the Red Sea, the Gulf of Aqaba, the Gulf of Suez and the Gulf of Aden. As with the Kuwait Convention, the Jeddah Convention focuses on oil pollution, both accidental and operational. Nutrient enrichment of marine eutrophication is not explicitly mentioned anywhere in this Convention. However, airborne forms of pollution as well as water borne forms of pollution from outfalls and pipelines (Article VI: Pollution from Land-Based Sources) are mentioned specifically. It is obvious that they have included pollution from nutrients and therefore eutrophication problems. Similar problems can arise from the exploration of the seabed and dredging.

3.5.11 Marine Environmental Protection of the South-East Pacific

The "*Convention for the Protection of the Marine Environment and Coastal Area of the South-East Pacific*" known as the "*Lima Convention*" was signed in 1981. The area of application of this Convention covers the "*sea and the coastal zone of the South-East Pacific within a 200-mile maritime of sovereignty and the high seas up to a distance within which pollution of the high seas may affect that area*" (Article I: Geographical Coverage). Hydrocarbons are especially mentioned in the Lima Convention; "*other*" harmful substances are also mentioned. There is no mentioning of marine eutrophication. However, the Lima Convention encourages the signatory parties to formulate, adopt and implement protocols for the protection of the marine environment against all types and sources of pollution (Article 3.4).

3.5.12 Protection of the South Pacific

The "Convention for the Protection of the Natural Resources and Environment of the South Pacific Region", known as the "Noumea Convention" was signed in 1986. Geographically the Convention covers a marine area extending up to 200 miles from the coast. There is no special mentioning of eutrophication within the text of the agreement.

3.5.13 The West Indian Ocean

The Convention for the Protection of the Western Indian Ocean, known as the Nairobi Convention was first signed in 1985, entering into force in 1996. This Convention with the full title "Convention for the Development, Protection, Management and Development of the Marine and Coastal Environment of the Western Indian Ocean" is part of the UN Environment's Regional Seas Programme. The Convention was amended in 2010 under the experience gained over thirty years of implementation. The Nairobi Convention was adopted in the first place by four contracting parties (UNEP/EAF, 2012): France, Madagascar, Somalia and Seychelles. This Convention was ratified eleven years later when Kenya and Tanzania signed it as well. Marine eutrophication is not referred explicitly in the Convention although large urban areas such as Mombasa, Dar es Salaam, Maputo, Durban Telaar produce considerable amounts of sewage ending in the marine environment.

3.6 Marine Environmental Protection in U.S.A.

There are three major laws in USA regarding the protection of the marine environment: (a) the Water Pollution Control Act (PL 92-500), known as the Clean Water Act (b) the Air Pollution Prevention and Control Act (PL 91-604) or Clean Air Act and

(c) the Coastal Zone Management Act (PL 92-583). All these three legal frameworks contain sections directly connected with nutrient discharges and over-enrichment of the marine environment, dealing therefore with pressures developing marine eutrophication. In addition to these federal laws, there are also state laws as well as regional programmes directly or indirectly associated with nutrient enrichment and marine eutrophication.

3.6.1 *The Clean Water Act*

The Jurisdiction of the water Pollution Control Act of Clean Water Act extends to all kinds of water bodies that is rivers, lakes and coastal waters. The main objective of the act is *"to restore and maintain the chemical, physical and biological integrity of the nation's waters"*. The basic philosophy behind the law is referred to as *"to follow the command-and-control method"* otherwise *"setting standards and enforcement"*. An innovative attitude endorsed in the 1987 revision was to discriminate between point-source pollution such as effluents from pipelines or outfalls and non-point sources of pollution. The reason for this distinction was that the contribution of pollution from non-point sources such as agricultural lands and urban sites has been underestimated. It is obvious that a great deal of municipal effluents and agricultural runoff were nutrient loads. The basic water quality criteria was set by EPA whereas the States were responsible for the development of standards ranging within the framework of the law as well as their implementation. Effluent limitations and water quality standards for water bodies were therefore, set by the States. As the Clean Water Act was not without critics, it was reauthorized during the 90s. The main recommendation by the National Research Council Committee on Watershed Management (NCR, 1990) was to adopt a policy that allows the *"bottom-up development of watershed agencies that respond to local problems rather than having a rigid institutional*

Table 3.2 Federal USA laws containing elements directly related to causes and effects of nutrient over-enrichment.

Federal law	Relevance to eutrophication
The Clean Water Act (Water Control Act, PL 92-500)	The main objective of the Act is *"to restore and maintain the chemical, physical and biological integrity of the nation's waters"*. The law refers explicitly to sewage pollution as well as pollution from agricultural activities and proposes policies to eliminate these forms of eutrophication.
The Clean Air Act (Air Pollution Prevention and Control Act, PL 91-604)	Although this Act was enacted to improve and maintain air quality in the United States, some aspects referring to atmospheric deposition in the marine environment indicate nitrogen and phosphorus enrichment from transboundary sources.
Coastal Zone Management Act (PL 92-583)	This act establishes a partnership scheme between The Federal and the State governments for coastal management. The latter includes regulations regarding nutrient over-enrichment and embraces many of the issues set by these three acts.

structure imposed upon them for the federal level". For this reason, the law encourages "*public participation in the development, revision, and enforcement of any regulation, standard, effluent limitation plan or program*". The administrator will draw minimum guidelines for public participation in these processes. It is also within the duties of the administrator, in collaboration with the competent authorities in agricultural activities to propose new and improved agricultural practices that would eliminate pollution from agriculture including nutrients. It is also among the administrator's duties to improve methods for "preventing, reducing, storing, treating or otherwise eliminating pollution" from rural areas where conventional methods of sewage management would be impracticable. In addition, wherever needed, sewage planning processes should be conducted and operated according to Section 201 of the Act.

3.6.2 The Clean Air Act

Air Pollution Prevention and Control Act (PL 91-604) commonly known as the Clean Air Act is a federal law with main objective to control air pollution in USA on a national level. However, in recent years, it has been realized that atmospheric deposition of nitrogen compounds (UNEPb, 2003) has a significant effect on the marine environment, especially in areas characterized as oligotrophic.

Amendments and revisions to the law have formed a powerful tool to reduce nitrogen over-enrichment of atmospheric origin in coastal and estuarine waters. It has been estimated by EPA (1998) that if the proposed standards by EPA had been implemented, the quantity of nitrogen of atmospheric origin would have had been reduced at least by 17 per cent. This has been the result of twelve case studies in estuarine environments. They had also calculated that the estimated "avoided costs" that is the costs to implement point source controls or storm water for the east coast estuaries would range between $150 and $250 million depending on the regulatory alternatives (EPA, 1998), if atmospheric controls were not implemented.

3.6.3 Coastal Zone Management Act

The Coastal Zone Management Act (1972) encourages the States to develop and implement coastal zone management policies within the objectives of the federal law. State policies should therefore define coastal boundaries, set standards, designate areas of concern and adopt enforceable policies covering the main objectives. The Law encourages "*the participation and co-operation of the public, state and local governments . . . . as well as Federal agencies having programmes affecting the Coastal Zone*". The law does not explicitly mention nutrient enrichment or eutrophication. However, nutrient-rich areas are recognized as the negative result of population growth and economic development. Nevertheless, it is obvious that any coastal zone management programme, takes into account the quality of the marine environment, expressed as marine water quality and ecosystem's health and this approach should take into account both eutrophic pressures and effects.

3.7 The European Union Policy in Marine Eutrophication

In most of the seas surrounding the European Continent, the Mediterranean Sea, the Black Sea, the Baltic Sea and the North Sea, the problem of eutrophication seems to be a priority in the agenda of most International Organizations. In addition, to the international conventions for the protection of the European Seas mentioned above, a number of EU Directives has been produced over the last thirty years for the protection of the different components of the marine environment. Some of them have included a component on marine eutrophication (Table 3.3). These Directives will be shortly presented below and their relevance to eutrophication will be enlightened.

Table 3.3 EU Directives on the marine environment and their relevance to eutrophication.

EU directive	Relevance to marine eutrophication
European Marine Strategy Directive MSFD (EC, 2008)	The MSFD requires, inter alia, from the Member States that *"human induced eutrophication is minimized, especially adverse effects thereof, such as losses in biodiversity ecosystem deficiency in bottom waters"*.
The Water Framework Directive WFDC (EC, 2000)	The problem of eutrophication is mentioned in the WFD Directive and it is required that nutrient concentrations should be measured. Reference values and nutrient scaling characterizing oligotrophic, mesotrophic and eutrophic conditions should be established. In addition, ecosystem quality can be expressed as *"composition abundance and biomass of phytoplankton"*.
Urban Waste Water Treatment Directive UWWT (EEC, 1991a)	This Directive suggests measures for mitigating eutrophication from *"accelerated growth of algae, undesirable disturbance to the balance of organisms and to the quality of the waters"* (Article 2§11). It also suggests identification of sensitive areas and treatment requirements: nutrient reduction is included referring to the removal of phosphorus and/or nitrogen.
Nitrate Directive (91/676/EEC0 (EEC, 1991b)	The whole Directive focuses on marine eutrophication from nitrates due to agricultural activities. Good agricultural practices are recommended. Annex III suggests measures to mitigate the problem.
Habitats Directive (92/43/EEC) (EEC, 1992)	There is no explicit reference on eutrophication. However, as the aim of the Directive is to ensure biodiversity conservation of natural habitats, the need for the protection of the aquatic systems, inter alia, from eutrophic pressures is obvious.
Bathing Water Directive BWD (2006/7/EC) (EC, 2006)	The objective of BWD is to ensure good coastal water quality for bathing and human health protection. However, bathing water profiles (Article 6) include assessment of phytoplankton and macro-algal growth (Annex III) which indicates the connection between BWD and the problem of eutrophication.

3.7.1 *The Urban Waste Water Treatment Directive*

The objective of the Urban Waste Water Treatment Directive (UWWT) is *"the collection, treatment and discharge of urban waste water aiming at the protection of the environment from adverse effects, results of waste water discharges"* (EC, 1991a). In eutrophic waters, nutrient reduction is required mainly by removing phosphorus and if possible nitrogen. Although eutrophication is not the only objective to be addressed within the UWWT, the contribution of the Directive towards the protection of coastal waters from *"accelerated growth of algae, undesirable to the balance of organism and to the quality of the water"* (Article 2§11) is significant. The same Directive provides a definition of eutrophication (Article 1.1) and also pays special attention towards sensitive areas and therefore treatment requirements (Article 5).

3.7.2 *The Nitrates Directive*

The objective of the Directive *"focuses on the Protection of Waters Against Pollution Caused by Nitrates from Agricultural Sources"* (EC, 1991b), otherwise known as the Nitrates Directive. The main objective of the Directive is the *"reducing of water pollution caused or induced by nitrates from agricultural sources"* and *"to prevent further such pollution"*. Article 2(i) provides a definition of eutrophication including *"undesirable disturbance"* whereas Article 6(c) explicitly refers to coastal eutrophication and the need for revisiting the situation every four years. The Directive establishes two management tools to mitigate nitrate pollution: (a) the designation of vulnerable zones (Article 3.4) and sensitive areas and (b) the development of good agricultural practices. These include limited use of fertilizers with appropriate (Annex II) measures as to prevent *"water pollution from run-off the downward movement beyond the reach of crop roots in irrigation systems"*. In addition, monitoring strategies are encouraged so that the success of the measures can be evaluated. The Directive does not specify any criteria for assessing levels of eutrophication and it is therefore necessary for Member States to set up scales of eutrophication (Karydis, 2009; Primpas and Karydis, 2011).

3.7.3 *The Habitats Directive*

The main objective of the Directive focuses "on the conservation of natural habitats and of wild fauna and flora" (EEC, 1992). The Habitat Directive shall be to "contribute towards ensuring bio-diversity through the conservation of natural habitats and of wild fauna and flora in the European territory of the Member States to which treaty applies" (Article 2.1). As loss of biodiversity in the marine environment is among the first symptoms when eutrophication trends appear (Gray, 1992), it is obvious that the problem of eutrophication is connected with the conservation of marine biodiversity. In the same article, Member States are encouraged to take measures to maintain or even restore natural habitats and species of Community interest (Article 2.2). The Directive is also encouraging Member States to carry out research and scientific work relevant to the objectives mentioned above.

3.7.4 The Shellfish Directive

This Directive is concerned with "*the quality of shellfish waters and applies to those coastal and brackish waters . . . . and thus to contribute to the high quality of shellfish products directly edible by man*" (EEC, 1979). Water quality assessment should be carried out regularly and the Annex of the Directive gives the parameters that should be measured. The directive does not mention eutrophication directly but some of the parameters like suspended matter and dissolved oxygen are closely related to eutrophic conditions. In addition, eutrophic conditions favor the development of toxic phytoplankton species (Ignatiades et al., 2007) as well as the survival of heterotrophic bacteria including pathogens (Karadanelli et al., 1992). This is the reason that the parameters saxitoxin and fecal coliforms have been included.

3.7.5 The Bathing Water Directive

The Bathing Water Directive (EC, 2006) lays down provisions for the management of the water quality for bathing and information to the public on bathing water quality. Although bathing water quality assessment is mainly focusing on a microbiological type of data (Annex I), the Directive is also concerned with the general condition of the water quality including inspections for materials such as tarty residues, plastic and rubber as well as the excessive growth of macro-algae and/or marine phytoplankton which is closely related with marine eutrophication (Article 9: other parameters). Although the Bathing Water Directive does not explicitly require an assessment of eutrophication, specific parameters such as water transparency, dissolved oxygen, nitrates and phosphates, have to be carried out every time that it is found that the quality of the water has deteriorated (ECOSTAT, 2004). Nevertheless, the Directive recommends "the wise use of resources" deriving from the Urban Water Treatment Directive and the Directive on "pollution caused by nitrates from agricultural sources" as well as the Directive in the field of water policy (EC, 2000).

3.7.6 The Water Framework Directive

The Water Framework Directive forms an integrated framework that aims to protect among other water bodies, transitional and coastal waters. The main objective is to ensure "*enhanced protection and improvement of the aquatic environment*" (EC, 2000). Amongst other requirements is a comprehensive ecological quality assessment for all waters, which describe the quality of the water masses on the basis of a number of biological, hydromorphological and physico-chemical quality elements, classifying the ecological status into five levels: "*high, good, moderate, poor and bad*" as compared to their deviation from specific reference conditions. Establishing reference conditions is one of the most significant stages in assessing ecosystem's status since the outcome is the result of comparison between observed values and reference conditions (Bald et al., 2005; Borja et al., 2012; Garmendia et al., 2013). The Good Ecological Status refers to the quality of biological communities (including phytoplankton, macrophytes and macrobenthos) where the ecosystem *"deviates only slightly from the conditions normally associated with the surface water body under undisturbed conditions"*. The community structure is also to discriminate between eutrophication levels (Arhonditsis et al., 2003). The critical point is always to define the boundary values between good and moderate conditions as they denote the early warning system for the data that have been collected. The Water Framework Directive requires from Member States to achieve at least a "Good Status" for their water bodies. Member States are also encouraged to develop marine quality indices that could be used for management purposes and policy making (Borja et al., 2007; Borja et al., 2006; Brito et al., 2012). There are therefore two prerequisites for the development of an eutrophication index (Karydis, 2009): (a) the establishment of reference conditions, that is values characterizing different variables of a marine community representing pristine conditions; these variables can be species composition, species abundance and biomass of phytoplankton (b) the development of a scale characterizing oligotrophic, mesotrophic and eutrophic conditions. This Directive places the emphasis on the state of the ecosystem and not on nutrient concentrations. Nutrient concentrations without detectable impacts are not a sufficient evidence for the downgrading of the Ecological Status. It is therefore necessary that research and monitoring (Andersen et al., 2006) should focus on the biological impact stemming from elevated nutrient levels and not merely from nutrient concentrations.

3.7.7 The European Marine Strategy Framework Directive

The overall aim of the Marine Strategy Framework Directive (MSFD) is to *"promote sustainable use of the seas and conserve marine ecosystems"* (EU, 2008). This objective will be achieved by protecting and preserving the marine environment by preventing its deterioration and *"where practicable, restore marine ecosystems in areas where they have been adversely affected"* (Article 1a). Like the Directive for water policy (EU, 2000), the MSFD is also oriented on the *"ecosystem-based approach"* ensuring that the pressures from human activities will be kept within limits that will assure good environmental status (Article 1c). The MSFD defines as environmental status *"the overall state of the environment in marine waters, taking into account the structure, function and processes of the constituent marine ecosystems"* (Article 3.4). The directive divides the European Waters into four main marine systems: (a) the Baltic Sea (b) the North-East Atlantic Ocean (c) the Mediterranean Sea and (d) the Black Sea adopting the ecoregion approach. This approach seems also to be suitable for eutrophication assessment and management as different European Seas differ in their trophic regime. The Central Area of the Mediterranean Sea is a typical oligotrophic environment whereas the Baltic Sea has always had elevated nutrient levels and phytoplankton biomass due to topographical and hydrological conditions of the system.

Member States will have the task to decide on a set of descriptors used to characterize good environmental status. Annex I of the MSFD indicates that the qualitative indicators for assessing human-induced eutrophication should be species biodiversity, harmful algal blooms and possible low oxygen values (hypoxia) in bottom waters. In addition, a description of the biological community including phytoplankton and its seasonal variability is needed (Annex III). A review on eutrophication indicators for assessing marine eutrophication within the requirements of the European Marine Strategy Framework directive has been published (Ferreira et al., 2011). Because of the pressures and impacts from fertilizers (nitrogen and phosphorus) due to agriculture as well as nutrient inputs from aquaculture, atmospheric deposition, sewers and riverine inputs should be quantified. A crucial point in MSFD is monitoring and assessment of the activities as monitoring will contribute to integration of conservation measures.

4

Eutrophication Status in the Regional Seas

4.1 Assessment and Effectiveness Criteria

The assessment of eutrophication in this section will have a twofold character: to evaluate marine water quality, including ecosystems' health, and assess the effectiveness of measures for mitigating eutrophic conditions set by the various Conventions for the regional seas. The latter enables decision and policy makers to consider, if necessary, possible reasons responsible for the deviation between objectives set and the real condition of the physical system.

Evaluations of the effectiveness of policies for marine environmental protection have been numerous. It starts from purely environmental issues and finally involves legal and policy aspects. An evaluation requires a number of criteria on which it will be based. There are several different criteria for assessing eutrophication (Karydis, 2009; Kitsiou and Karydis, 2011; Ferreira et al., 2011) as well as a number of methodological approaches already presented in Chapter Two (methodology) of this book. The effectiveness regarding environmental governance using various criteria has also been a subject of study by various authors: Andersen and Wettestad (1995), DiMento (2003), Haas et al. (1993), Keohane and Levy (1996), Miles and Underdal (2001). They expressed different approaches but they all agreed that effectiveness is closely connected with solutions of environmental problems that involved cooperation among states, international organizations, the scientific estate, non-state actors and the participation of the public. As regional seas are of invaluable importance to society and the nature itself, their governance is of growing interest with more political issues emerging with time. A number of indicators of marine environmental quality regarding eutrophication problems and indicators of effectiveness have been selected as the basis of the assessment of eutrophication from the physical and legal point of view for the present Chapter. By the term indicator we mean a number that provides information to an issue or a condition. Effective indicators are objective ways to assess a situation regarding marine water quality, ecosystems' health, trends in the society or effectiveness of economic measures and policies. The advantage of an indicator is that it is based on a subset of the accessible data and it is easy to produce; there are two points, however, that should be taken into account when an indicator is selected (a) varies according to the types of systems they monitor. This is why the EU Directives, namely the Water Framework Directive and the Marine Strategy Directive encourage Member States to develop specific indicators on ecosystems' health for each ecoregion and (b) there are four criteria when selecting an indicator, to ensure suitability for the information they provide. Relevance is the first feature that an effective indicator should fulfill. That means they have to show the aspect of the system we are looking for. The second feature is that they must be easy to understand. As their information is addressed to a range of scientific disciplines, that is, scientists, politicians, sociologists and economists, the indicator should be understood by groups of experts beyond the group it comes from. The reliability of an indicator is another criterion as important as the criteria mentioned above: the effectiveness indicator should be relatively sensitive to changes but at the same time robust to secondary or random fluctuations such as seasonal trends; the last criterion refers to the accessibility of data: data are not always accessible especially in the marine environment. In many cases they are not accessible in real time. The selected indicators and a brief description referring to their necessity is given in Table 4.1.

The effectiveness in mitigating discharges is a legal component in many marine regional Conventions. Protocols on pollution on land-based sources refer to outfalls and pipelines. They also refer to sewage, anthropogenic source of phosphates and nitrates as well as to the need for sewage treatment units. Another source of pollution favoring eutrophic conditions is the non-point enrichment of the marine environment

Table 4.1 Indicators of water quality and effectiveness indicators of environmental governance used in the assessment of eutrophication in the regional seas.

Indicator	Description
Effectiveness in mitigating discharges	Most of the Conventions refer to specific protocols on discharges of sewage and industrial effluents into the marine environment.
Effectiveness in protecting biodiversity	Many of the international Conventions for the protection of the marine environment require measures for biodiversity conservation.
Effectiveness in keeping good ecosystems' health	The emphasis in many legal documents such as EU Directives is placed on the health of marine ecosystems.
Effectiveness in minimizing extreme events	Intense eutrophic conditions in combination with their physical and chemical factors can trigger algal blooms sometimes dominated by toxic marine species.
Effectiveness in contributions to relevant legal instruments	Compliance regarding international obligations, national measures, marine environmental monitoring, transboundary eutrophication assessment impacts and environmental management should be evaluated.
Effectiveness indicators describing good collaboration among states and people	International cooperation among states, institutions, non-state factors, improves the quality of their relations leading to measurable changes in upgrading environmental quality including improvements of the trophic conditions.

from fertilizers and animal farming. The second indicator is the assessment of biodiversity. Biodiversity is also a component in many Conventions (OSPAR, 2005; Karydis and Kitsiou, 2014) and EU Directives (EC, 2000; EC, 2008; EEC, 1992) and closely connected with good ecosystems' health. The main threat in biodiversity is nutrient pollution and eutrophication, causing consecutive impacts that in extreme conditions can even lead to anoxia. These impacts have been presented in the Section 1.4 of this book.

The effectiveness in keeping good ecosystems' health is explicitly required in two EU Directives: The Water Framework Directive and the Marine Strategy Directive. It also appears in many Conventions regarding regional seas. The effectiveness in minimizing extreme events is also a priority closely connected with marine eutrophication toxic algal blooms and mass fish mortality; they are both related to eutrophic conditions.

An Indicator of effectiveness adopted in the present work refers to contributions to relevant legal instruments. These can be obligations, national measures, marine environmental impact assessments, transboundary pollution issues and environmental management (DiMento and Hickman, 2012). Possible legal instruments developed and the quality of implementation in relevance to marine eutrophication problem is discussed wherever possible. The Effectiveness Indicator by good collaborations among states and people is also very important for achieving good marine environmental quality objectives. It has been supported that measurable improvements in coral reefs, decrease of eutrophication levels, oil pollution, just to mention a few, need close collaboration among nation states (DiMento and Hickman, 2012). This means that the literature from marine environmental sciences, social sciences, economy and law will be taken into account. Although the objective is marine eutrophication (present and future perspectives), eutrophication issues can be delineated better in a holistic system that is within the social, economic, political and legal framework. This is going to be the approach in the following section, presenting marine eutrophication issues in the regional seas.

4.2 Eutrophication Status in the Mediterranean Sea

4.2.1 Physiography of the Mediterranean Sea

The Mediterranean Sea is a semi-enclosed basin connected to the Atlantic by the Strait of Gibraltar, 15 km wide and 290 m sill depth; to the Red Sea through the Suez Canal and to the Black Sea through the Straits of Dardanelles, having a maximum width of 7 km and an average depth of 55 m. The Mediterranean Sea covers an area of approximately 2,500,000 km^2. The length of the basin from west to east is about 3,800 km and the maximum north to south distance is about 900 km (Karydis and Kitsiou, 2012). The Mediterranean Basin is divided into two main sub-basins, the western sub-basin and the eastern sub-basin which are separated by the Sicilian Channel; the width of the Sicilian Channel is having a width of about 150 km and 400 m maximum depth. The bottom of the Mediterranean Sea is relatively flat with an average depth of about 2,700 m. On the contrary, the eastern part is characterized by steep slopes, submarine valleys and alternating depressions and high morphological structures. The maximum depth of the Eastern Mediterranean is 4,982 m, located off the shores of the south-west Greece in the area of Peloponnese. On the other hand, the shallowest system of the Eastern Mediterranean is the Northern Adriatic Sea where the sea depth does not exceed 200 m. The western sub-basin has been divided into four marine areas (Cruzado, 1985; UNEP, 2003) illustrated in Figure 4.1: I. The Alboran Sea surrounded by Algeria, Morocco and Spain, II. The Northwestern Basin surrounded by Italy, Monaco, France and Spain; III. The Southwestern Basin surrounded by Tunisia, Algeria, Spain and Italy and IV. The Tyrrhenian Sea surrounded by Italy, France and Tunisia. The eastern sub-basin is divided into six marine areas: V. The Adriatic Sea, surrounded by Albania, Croatia and Italy, VI. The Ionian Sea surrounded by Greece, Albania and Italy, VII. The Central basin surrounded Italy, Greece, Libya, Tunisia and Malta, VIII. The Aegean Sea surrounded by Turkey and Greece, IX. The North Levantine Sea surrounded by Lebanon, Syria, Cyprus, Turkey and Greece and X. The South Levantine Sea surrounded by Egypt, Libya, Israel and Lebanon.

The big rivers outflowing in the Mediterranean are only few, the River Nile in Egypt being the greatest; the catchment area of Nile extends well inside the NE African continent by several thousands of kilometers. The length of Nile is 4,132 km, the catchment area 3,350 million km^2 and the water outflow about 90 km^{3}y^{-1}. Rhone is the second river in the Mediterranean regarding the water outflow; it rises in the Central Alps in Switzerland and ends in the Gulf of Lion (France). The third most important river is the Po River, draining the southern flanks of the Alps and outflowing in the Adriatic Sea. The fourth in size Mediterranean river is the Ebro River crossing Spain. The catchment areas of Rhone, Ebro and Po rivers are 96,000, 84,000 and 69,000 km^2 respectively, whereas their outflow is 54, 17 and 46 km^{3}y^{-1}.

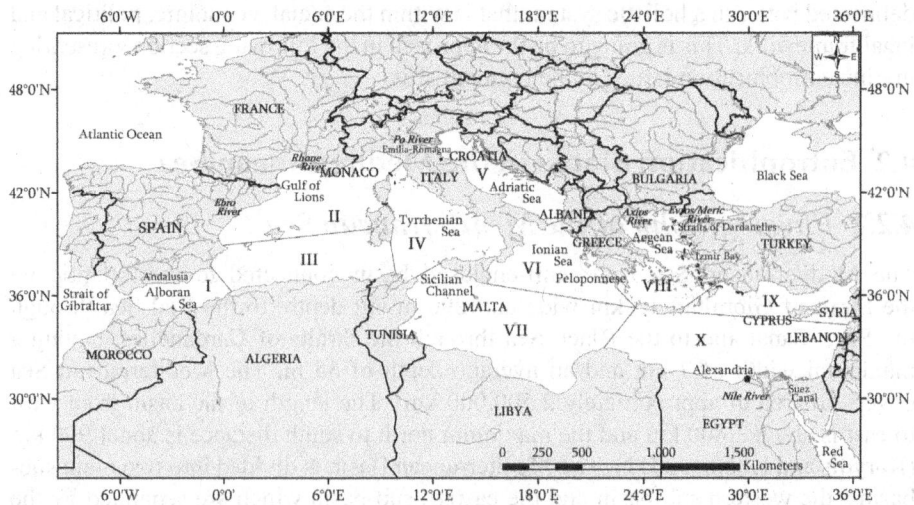

Figure 4.1 The Mediterranean Sea divided into ten sub-basins: I Alboran Sea, II Northwestern Basin, III Southwestern basin, IV Tyrrhenian Sea, V Adriatic Sea, VI Ionian Sea, VII Central basin, VIII Aegean Sea, IX North Levantine Sea, X South Levantine Sea. Source EEA (1999)-modified.

The Mediterranean climate ranges from subtropical to mid-latitude weather systems (EEA, 1999). The main characteristics are windy, mild wet winters and relatively calm, hot and dry summers. The rain falls from late autumn up to early spring, accounting for more than 90% of the yearly precipitation. There is a tendency for rainfall to decrease southwards but the geographical distribution is also influenced to a significant degree by the orography.

The water structure of the Mediterranean Sea seems to be comprised of three major water masses: (a) the Atlantic water inflowing as surface water into the Mediterranean, the inflow-outflow balance being 1,700 km^3y^{-1}, (b) the Levantine Intermediate Water and (c) the Mediterranean Deep Water (Bethoux, 1980; Bethoux et al., 1998). The circulation pattern is rather complex but the cyclonic pattern seems to predominate (POEM Group, 1992).

4.2.2 Environmental pressures and sea conditions of the Mediterranean Sea

The main environmental pressures in the Mediterranean Sea stem from the growth of population: the population of coastal states in 1960 was 246 million, escalating to 380 million in 1990 and 250 million in 1997. Now it seems to exceed the number of half a billion people. One third of the Mediterranean population is concentrated in coastal areas. All these people and their activities along the coastal zone exercise a number of pressures on the marine environment. The main pressures come from tourism, agriculture, aquaculture industry and discharges from sewage outfalls (EEA, 1999).

After the Second World War, the phenomenon of mass-tourism started to develop along many Mediterranean seaside areas. These days the Mediterranean is the biggest holiday resort in the world; it accounts for 30 per cent of the international tourist arrivals and 25 per cent of the gross income from international tourism. According to a Blue Plan scenario the number of tourists in the coastal areas of the Mediterranean will be as high as 350 million tourists by the year 2025. The Mediterranean tourism is characterized by heavy seasonality, high tourist concentration along coastal areas and a preference of the tourist market to the north-western Mediterranean. The overall number of inhabitants in the coastal zones (both locals and tourists) exercises pressures on land use, consumption of fresh water resources, pollution and waste disposal, enhancing this way the problem of eutrophication. On the other hand, tourism requires good environmental quality in both the terrestrial and the marine environment and therefore it can exercise a counter pressure for environmental protection and implementation of best practices in coastal management policy.

Agricultural practices include, *inter alia*, plant cultivation, irrigation, dairy farming and pasture; all these activities cause various non-point forms of pollution. From the point of view of eutrophication, fertilizing leads to nutrient run-off, especially nitrates and phosphates leading to eutrophication. Manure spreading also leads to phosphate and nitrate pollution. Suspended matter is also carrier of nutrients, contributing to eutrophication once it is discharged into the sea. There are also other forms of pollution not directly connected with marine eutrophication such as pesticides and pathogens.

EU Member States like France, Italy, Spain and Greece have to comply with the Nitrate Directive and make reasonable use of fertilizers. However, many non EU Mediterranean States such as Egypt, Israel, Lebanon, Syria and Turkey make intensive use of fertilizers also causing transboundary marine water pollution (UNEP, 2003b). Large river basins discharge considerable amounts of water and therefore nutrients: the highest discharges come from the rivers Nile (Egypt), Po (Italy), Rhone (France), Ebro (Spain), Evros/Meric (Greece/Turkey) and Axios (Greece). In addition, there is airborne nutrient transport mainly from the Sahara Desert (UNEP, 2003b; Morales-Baquero et al., 2013).

Marine aquaculture production in the Mediterranean follows an exponential trend: 78,180 tons in 1984 and 248,460 tons in 1996. In 2001, it exceeded 1,349,777 tons and the trend is still upward (Basurco and Lovatelli, 2016). The marine aquaculture production since 1984 has increased at least by 400-fold. The interaction of aquaculture with the marine environment is characterized by many negative effects. Production of waste can stimulate and/or distort productivity and induce changes of the biotic and abiotic characteristics of the sea but the main problem is the hyper-nutrification leading to the establishment of eutrophic conditions. This can also affect the seabed conditions, causing oxygen depletion and change the ecological characteristics of the site. Ackeforsand and Enell (1990) have proposed a simple formula for estimating the total quantity of nitrogen and phosphorus entering the marine environment in Swedish fish farming; the formula depends on the Food Conversion Ratio and the fish biomass produced. This formula has been adapted for all countries by Ceccarelli and Di Bitetto (1996). Estimated loads of phosphorus and nitrogen from marine and diadromus fish in the marine environment for many Mediterranean States between 1984 and 1996 have been given in the EEA Report (1999).

Oil and natural gas are produced in many Mediterranean countries (Libya, Algeria, Tunisia, Egypt and Syria) leading to the establishment of oil terminals, liquidification infrastructures for natural gas and many refineries along the Mediterranean basin (Kitsiou and Karydis, 2014). There is also production of mercury, chromium, lead, salt, boxite and phosphates just to mention a few. Apart from the phosphates, the rest of the products, although highly toxic to marine organisms, are not closely connected to eutrophication.

The MED POL Programme, although is a multi-objective programme has paid special attention on the "*Survey on Land-Based Sources of Pollution*" (WHO/UNEP, 1996) and the "*Identification of Priority Pollution of Hot Spots in the Mediterranean*" (WHO/UNEP, 1997). However, they are both focusing on the sewage effluents mainly discharged from big Mediterranean coastal cities. Not all coastal cities have built facilities for sewage treatment; in some cases, waters from treated sewage are reused for irrigational purposes. In addition to coastal cities, river discharges are another source of nutrients in the marine environment. Although nutrient levels in the Mediterranean Sea are on average four times lower than the rivers crossing Western European countries, their contribution to marine eutrophication in coastal waters is sizable. However, with the only exception of the Adriatic Sea that has been characterized as eutrophic (Karydis and Kitsiou, 2012) and various coastal areas, the main water body of the Mediterranean Sea is typically oligotrophic.

4.2.3 Eutrophic conditions in the Mediterranean Sea

The Mediterranean Sea has been characterized, as mentioned above, as a typical oligotrophic system, may be the most oligotrophic system in the world (Ignatiades et al., 2009; Karydis and Kitsiou, 2014; UNEP, 1989). The oligotrophic character of the Mediterranean can be explained on the basis that water inputs through the surface layer in the Strait of Gibraltar are poor in nutrients compared to the outflowing water mass through the deeper layers of the Gibraltar water masses since the estimated ratio of outflowing nitrogen/inflowing nitrogen is $5.7/1.9 = 3.1$ (UNEP/FAO/WHO, 1996). Calculations in the inflow–outflow nutrient budget indicate that the Atlantic surface water accounts for 71 per cent of the deep water outflow. The rest (29 per cent) is due to terrestrial input (UNEP/FAO/WHO, 1996). However, nutrient concentrations in deep offshore waters (> 1,000–2,000 m) are of no interest in eutrophication processes since these depths are well below the euphotic zone. An exception can be the upwellings but these occur only in some coastal areas and enhance eutrophic conditions only at a local scale.

It is known that oligotrophy is the result of nutrient limitations in the surface waters but there is a debate regarding the limiting factor: nitrogen or phosphorus? Krom et al. (1991) have supported the view that the Levantine basin is limited by phosphorus. Similar views about the limiting nutrient have been expressed for the Adriatic Sea, a sea considered for the most of its area as eutrophic. Phosphorus seems to be the limiting factor along the coastal waters of Emilia–Romagna (Mingazzini et al., 1992). On the other hand, data from surface waters from the Ligurian Sea indicated a slight tendency for nitrogen to be the limiting nutrient (Bethoux et al., 1992). Nutrient and chlorophyll α concentration values from the various Mediterranean sub-basins with their relevant references are given in a review by Karydis and Kitsiou (2012). The average annual productivity in the Eastern Mediterranean Sea (EMS) has been estimated by Psarra et al. (2000) to be between 60 and 80 $gCm^{-2}y^{-1}$ which is almost half compared to the productivity of other oligotrophic areas (Krom et al., 2010) such as Sargasso Sea (125 $gCm^{-2}y^{-1}$) and the Northeast Pacific (120–130 $gCm^{-2}y^{-1}$). On the contrary, primary productivity in the Eastern Ionian has been found to be 285.26 $gCm^{-2}y^{-1}$ (Pagou and Gotsis-Skretas, 1990) and in the Gulf of Lions 140–150 $gCm^{-2}y^{-1}$ (Lefevre et al., 1997). More information on the Mediterranean primary productivity can be found in the work by Ignatiades (1990); Karydis and Kitsiou (2012).

Eutrophic areas in the Mediterranean seem to be limited to specific coastal systems and adjacent offshore areas, characterized by eutrophication events. This is the case near big cities, semi-enclosed bays and some estuaries. Aquaculture is also causing eutrophic conditions, usually at a local scale. The northern shores of the Mediterranean seem to be the most affected and serious eutrophication problems have been documented. However, the situation in the south should not be underestimated: the limited number of eutrophic events may be the result of poor monitoring. In any case the development of the North African States, the increase of mass tourism and the establishment of industrial facilities may be a future threat for marine eutrophication in the Southern Mediterranean (Garmendia et al., 2015).

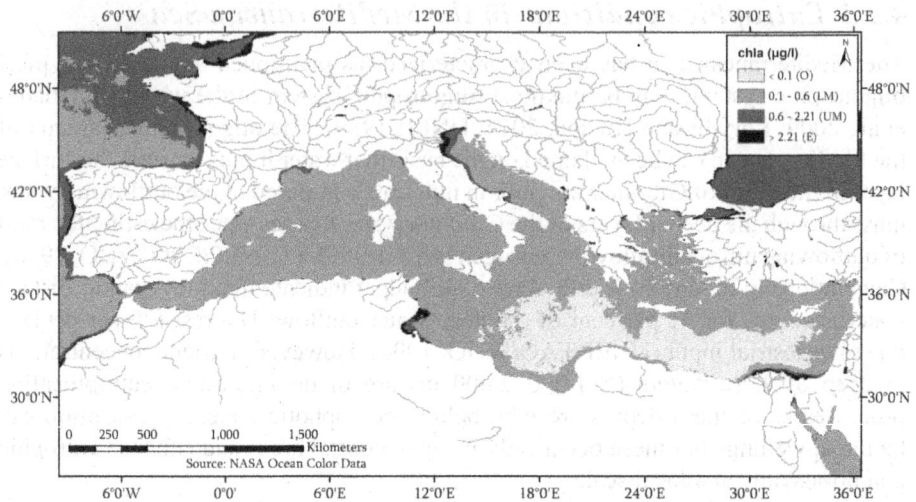

Figure 4.2 Satellite surface chlorophyll distribution in the Mediterranean Sea during 2017. Eutrophication scaling according to Simboura et al. (2005-modified) scale.

Surface chlorophyll distribution in the Mediterranean Sea is shown in Figure 4.2. Eutrophic areas are located in the Northern Adriatic Sea and the eastern coastal areas of the Italian peninsula. Coastal eutrophication is also observed in the Gulf of Lions as well as along the coastal areas of Spain between Almeria and Gibraltar. In the Southern part of the Mediterranean, eutrophic areas are located in the Gulf of Gabes (Tunisia) and the coastal zone between Alexandria and Port Said. It must be noted that most of the area of the Medditerranean Basin is classified as lower mesotrophic, whereas an area covering the SE part of the basin is oligotrophic.

In addition to riverine effluents estimated between 400 and 450 km^3y^{-1} (total riverine effluents), fluxes from agricultural activities have to be taken into account. Nitrogen used in agriculture varied between 30 and 70 $kgNha^{-1}$. High nitrate fluxes have been found in the Adriatic Sea and were estimated about 250,000 ty^{-1} whereas nitrogen fluxes in the Western Mediterranean (between Genoa and Valencia) were 340,000 ty^{-1} (Vukadin, 1992). The main source of phosphorus is the urban wastewater (GPA, 1995). The increase of phosphorus fluxes in the Mediterranean Sea was found to be twice as high as the increase of nitrate fluxes (Cruzet et al., 1999). Phosphorus loads in the Adriatic Sea ranged between 82,000 and 115,000 ty^{-1} for the western arc (Vukadin, 1992).

Marine eutrophication is not accessed only by nutrient and chlα concentrations but also by the frequency of algal bloom events, sometimes dominated by toxic marine phytoplankton. Significant quantities of *Gymnodinium catenatum* have been reported in the Andalusian coastal area by Bravo et al. (1990). Delgado et al. (1990) has found a large bloom of *Alexandrium minutum* in the Ebro Delta which is a PSP toxin flagellate during May 1989. Belin et al. (1989) and Lassus et al. (1991) have also reported the presence of HABs along the French Mediterranean coasts; *Alexandrium minutum* and *Dinophysis* sp., were the dominant species. Algal concentrations as high as 6,000,000 cells l^{-1} of the toxic species *Gymnodinium catenatum* were found at Fysano lagoon near Naples (Carrada et al., 1988). Frequent algal blooms have occurred in the Adriatic Sea near the Emilia-Romagna coastal area (Boni et al., 1992). Toxic phytoplankton has also been identified in many Gulfs in the Aegean Sea. *Alexandrium minutum* had the most frequent occurrence (Ignatiades et al., 2007; Ignatiades and Gotsi-Skretas, 2010). Similar occurrences have been reported in the Izmir Bay, in Alexandria (Egypt) as well as in the coastal area of Libya, Tunisia, Algeria and Morocco, although the information is scanty. More information about toxic phytoplankton and algal blooms in the Mediterranean Sea can be found in EEA (1999) and Karydis and Kitsiou (2012).

Airborne nutrient deposition in the Mediterranean Sea seems to play a significant role due to the oligotrophic character of the area. There are indications (Morales-Bequeto et al., 2013) that dust from the Sahara Desert increases the total phosphorus in the SW Mediterranean. This view has been proposed in an earlier work by Krom et al. (2004). However, the significance of atmospheric deposition in the marine environment was understood over the last few years and there are still difficulties to quantify atmospheric deposition nutrient processes (Moon et al., 2016). According to the UNEP's Report on Transboundary Pollution (UNEP, 2003b) concentrations of NO_3^- and NH_4^+ in the Mediterranean rainwater were 36 ± 20 and 30 ± 11 μeql^{-1} respectively (average values). However, information on gaseous forms of HNO_3^- and NH_3 is limited in the Mediterranean. Emissions of NO_x, NH_3 and total N over the Mediterranean Sea have been estimated by Vestreng and Klein (2002): they were 1,800, 2,300 and 4,200 kt respectively of nitrogen during 1999. A more detailed nutrient deposition study in the Mediterranean for each of the ten areas (Figure 4.1) has been given by Tarrason et al. (2000). On the contrary, calculations on phosphorus depositions in the Mediterranean are rather limited. Atmospheric inputs of phosphorus into the sea were estimated about 40 mg Pm^{-3}y^{-1} in the Eastern part of the Mediterranean (Bergametti et al., 1992).

4.2.4 Policy effectiveness in the Mediterranean Sea

Having presented the Legal Framework (Chapter Three) and the conditions of the sea (Section 4.2.2), an evaluation of the effectiveness of policy in the Mediterranean Sea will be attempted by taking into account as many of the indicators presented at the beginning of Chapter Two as possible. It has to be mentioned that in spite of the many different forms of pollution in the Mediterranean, eutrophication, oil pollution and emerging pollutants are the main problems (Danovaro, 2003). The emphasis in the current section, however, will be placed on eutrophication.

As a general outcome, it can be said that there are encouraging trends. The UNEP/MAP-Plan Bleu Report (2009) as early as 2009 concluded that *"Conservation actions implemented to date had positive results and some species have already been saved from extinction. Namely from 2001 to date 175 large biodiversity projects have been identified within the Mediterranean Region"*. This is also an indirect indication that the mitigation of eutrophic conditions has succeeded to a degree. Although it is difficult to have the exact information on sewage treatment around the Mediterranean, there are data on the water quality of swimming beaches that indicate an improvement in the marine water quality: monitoring covering more and more beaches within the MED POL Framework, has shown that the number of beaches that comply with the required standards increases steadily. This is an indication "*that treatment and disposal of sewage are improving in many parts of the Mediterranean*" (Keckes, 1994). A good example is the coastal zone of France in the Mediterranean Sea. It extends up to 1,700 km and the permanent population is 5.8 million permanent residents whereas the annual number of tourists is about 29.2 million. More than 93 per cent of the population in the area is covered by public sewage systems (Keckes, 1994). It has been concluded that "*the status of the Mediterranean, considering multiple impacts, is still in a pretty good shape after more than 25 years of the programme. When some activities were not the responsibility of UNEP/MAP, such as fisheries, the results are not so good. Another important point is the awareness that has not been built in all sectors involved with environmental matters*". However, it is difficult to draw definite conclusions in the Mediterranean environment and the quality of indicators used. Efforts have still to be made to reduce pollution from land based sources, to adopt the ecosystem approach as a quality measure to establish a complete network of marine protected areas and work out plans for integrated coastal management.

The Barcelona System is a "*model Convention*" and it has been used as a platform for the Conventions in other regional seas for the protection of the marine environment under the aegis of UNEP. In addition, the system is adaptable to developments of the international law for marine environmental protection. At the phase of implementation, a number of tools have been adopted so as to provide scientific evidence regarding the effectiveness of the measures proposed in the Barcelona Protocols. A powerful tool promoted by the Barcelona System is "*Environmental Impact Assessment*" (EIA). EIA are a standard requirement in most Mediterranean countries for major development projects that may have complex implications (OECD, 1996). The "*Eutrophication Monitoring Strategy of MED POL*" launched in 2003 is one more tool for assessing eutrophic conditions in a number of Mediterranean States that have recorded cases of eutrophication in hot spots, river estuaries and semi-enclosed bays. This strategy plan that covers parameters to be monitored, sampling frequency, spatial coverage and data quality assurance places the emphasis of the Barcelona System on eutrophic issues (UNEP/MAP, 2007). Many scientific articles have been published since then on eutrophication of the Mediterranean (Siokou-Frangou et al. 2010, Karydis and Kitsiou, 2010, 2012; Garmendia et al., 2015).

A shortcoming regarding the effectiveness in the present case is the lack of compliance. Compliance in the Barcelona System is restricted to the submission of reports to UNEP. As a result, there is no uniform concept regarding the implementation of the requirements set at the protocols.

The improvement in the relations among states can be considered as a positive outcome from the functioning of this Convention. It is a good example of cooperation among states "*characterized by different political systems, religions, wealth development and resources among its members*" (DiMento and Hickman, 2012). This is particularly important since various forms of marine pollution including eutrophication are of transboundary nature and a form of agreement among the states involved in a particular eutrophication problem is necessary. An example about the transboundary nature of eutrophication can be rivers crossing many countries, carrying nutrients (or sewage and particulate material of agricultural origin) to the estuarine environment. Airborne nutrient deposition is another example of transboundary nature. From this point of view, the Mediterranean Action Plan has been characterized as a coordinating mechanism recognized by everybody (DiMento and Hickman, 2012).

In conclusion, according to DiMento and Hickman (2012), the Mediterranean Action Plan is the oldest and at the same time among the most developed regional seas programmes and possesses a comprehensive cluster of governance components. It seems to be an effective platform for marine environmental affairs in spite of the political turmoil in the areas over the last few years and financial problems of the numerous Member States. The contribution of this scheme to the mitigation of eutrophic problems is obvious and well documented.

4.3 Eutrophication Status of the Black Sea

4.3.1 *Physiography of the Black Sea*

The Black Sea is the world's largest inland sea; it is located among Eastern Europe and Caucasus, Western Asia and surrounded by Russia, Turkey, Ukraine, Georgia, Romania and Bulgaria. The Black Sea area is about 436,000 km^2; this number does not include the area of the Azov Sea which is 38,000 km^2. The Black Sea volume is 547,000 km^3, the maximum depth 2,212 m and the average depth 1,240 m (Zenkevich, 1963; Ross, 1977). The catchment area is 1,874,904 km^2 (Vespremeanu and Golumbeanu, 2018) and 23 countries are completely or partially contained in the catchment area (UNEP, 2006).

The only communication of the Black Sea with the world's seas is through the connection with the Mediterranean (Aegean Sea) and finally to the World's Ocean via a number of Straits (Figure 4.3). The Bosporus Strait connects the Black Sea to the Sea of Marmara and the Strait of Dardanelles connects the Marmara Sea with the NE Aegean Sea. The Strait of Bosporus is among the narrowest Straits in the World, having a total length of about 3 km and an average width 1.6 km; the average depth is about 36 m.

The Danube, Dnieper and Don rivers which are the three out of the four largest rivers in Europe outflow into the Black Sea. The total annual outflow of these rivers has been estimated to be 350 km^3. Due to this excessive fresh water input and the fact that the Black Sea is a positive sea regarding the water balance, there is an outflowing water mass into the Northern Aegean Sea. There is also an inflowing bottom current from the Mediterranean but the outflowing volume is twice as much as the inflowing. The two currents are characterized by high speeds allowing a turbulent mixing of the two layers. The salinity of the outflowing water mass is 17 psu whereas the salinity of the Mediterranean inflowing water mass is about 34 psu.

The Black Sea's upper low salinity layer receives oxygen from the atmosphere. On the contrary, deep waters do not mix with the surface layer and are therefore characterized by anoxic conditions: 90 per cent of its water volume is anoxic. The circulation pattern is mainly controlled by the basin topography and fluvial inputs; this explains the strong vertical stratification structure. Due to this extreme stratification,

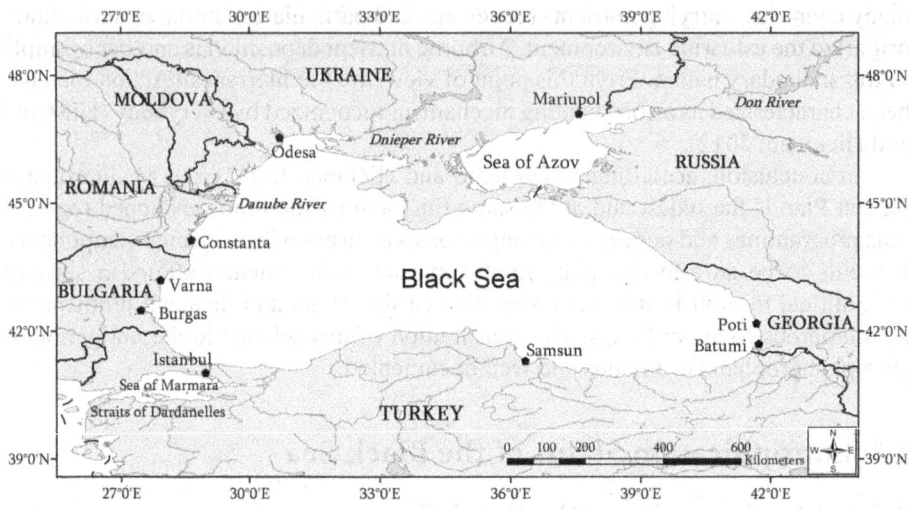

Figure 4.3 The Black Sea.

the Black Sea is classified as a salt wedge estuary. The structure of the water body mentioned above is closely connected with the ecology of the basin as well as with pollution problems mainly due to the semi-enclosed character of the basin. The general surface circulation pattern is cyclonic. Two large cyclonic gyres have been identified: one gyre in the western part and another in the eastern part. Although this general circulation pattern has been known since the 19th century, the driving forces that cause this circulation are not known (Sur et al., 1996; Ivanov and Belokopytov, 2013). However, two factors have been considered of principal importance: the wind for the surface circulation and river runoff (Moskalenko, 1976; Marchut et al., 1975).

Atmospheric conditions show a wide range of temporal and spatial variability. This variability is observed from east to west and from north to south. This is because the climate of the area is affected by the Mediterranean climate as well as the steppe climate. During winter, the center of maximum pressure prevailing in Siberia dominates the region. On the other hand, the Azores anticyclones can disturb atmospheric conditions during the October to March period, leading to a temperature drop, precipitation and strong winds. Most of the depressions in the area are accompanied by fronts (Balkas et al., 1990).

4.3.2 *Environmental pressures and sea conditions in the Black Sea*

In the past, human pressures have transformed the Black Sea from a healthy ecosystem characterized by high diversity into a eutrophic marine system (Konovalov, 1995). The coastal population of the countries bordering the Black Sea that is Romania, Ukraine, the Russian Federation, Georgia, Turkey and Bulgaria is about 17.5 million permanent inhabitants but to this number 6 to 8 million tourists per year have to be added (Vespremeanu and Golumbeanu, 2018). The tourism industry started growing after the Cold War as many coastal areas became very popular tourism spots. Although economic benefits from this source of income are significant for the riparian states, these activities exercise extra pressures on the coastal and marine system (BSC, 2009). The coastal population is unevenly distributed among the riparian countries. There are 63 harbors around the Black Sea perimeter but the most important are Constanta (Romania), Odessa (Russia), Mariupol (Ukraine), Poti and Batumi (Georgia), Samsun and Istanbul (Turkey), Burgas and Varna (Bulgaria). Tourism around the Black Sea is more developed in Bulgaria, Romania and Ukraine.

River runoff into the Black Sea accounts for most of the nutrient discharges into the basin. The outflow of Danube, Dniester and Dnieper rivers into the Black Sea is about 350 km^3y^{-1}. The Danube River is the biggest accounting for 250 km^3y^{-1} of water outflowing which is approximately 75 per cent of the total fresh water input. The Dnieper and Dniester rivers outflow 53 km^3y^{-1} and 9.6 km^3y^{-1} respectively. River runoff is the most important nutrient enrichment (Konovalov et al., 1999; Konovalov and Murray, 2001; Ludwig et al., 2009; Strokal and Kroeze, 2013; Mikealyan et al., 2015, 2017). The inflowing water volume exercises dynamic control over the water balance of the basin affecting circulation and environmental conditions. Furthermore, river outflow stabilizes water stratification. The structure of water masses limits vertical movements. The development of a narrow current produces a 'scatter' of nutrients over the entire sea surface (Yankovsky et al., 2004).

Inflowing nutrient volumes have been calculated by various authors: nitrogen fluxes from the catchment area of the Danube River showed an increase from about 400,000 ty^{-1} in the 1950s to about 900,000 ty^{-1} during 1985–1995 (daNUbs, 2005). This nutrient outbreak showed a decline over the next decade. It was reduced to 760,000 ty^{-1} during 2000–2005 (Pakhomova et al., 2014). River outflows are transported along the coastal areas through turbulent exchanges and filament structures, ending into the inertia of the basin (Yunev et al., 2002).

The phosphorus fluxes were minor compared to nitrogen but also with an upward trend: the 40,000 ty^{-1} of phosphorus released in 1950s reached a peak of 115,000 ty^{-1} at the beginning of 1990s. The flux also showed a decrease during 2000–2005 (70,000 ty^{-1}). The construction of a dam and a reservoir by the Iron Gate had a minor influence in the nutrient discharges (daNUbs, 2005). A combination of factors, i.e., anthropogenic impacts, climate changes and hydrological oscillations have changed the ecosystem regime of the basin (Kideys, 2002).

Apart from the river nutrients, the basin is suffering from chemical pollution, urban and industrial waste disposal, overfishing and an alien predator *Mnemiopsis leidyi*. Heavy metals brought into the Black Sea environment by the rivers contribute significantly to water pollution. The quantities of heavy metals discharged into the sea form a serious threat: 1,400–1,500 ty^{-1} of copper, 60 ty^{-1} of mercury, 280 ty^{-1} of cadmium and 6,000 ty^{-1} of zinc are only among the most important pollutants to be mentioned (Vespremeanu and Golumbeanu, 2018). Out of these quantities, 55 ty^{-1} of mercury and 240 ty^{-1} of cadmium are released by the Danube River (Konovalov, 1995). The organochlorine pesticides is another form of chemical pollution in the Black Sea. The main way of transportation of these compounds is the wind blowing from agricultural land and forests. Maximum pollution rates from DDT and chlorinated hydrocarbons were recorded during the 1960s and 1970s when agricultural practices were rather poor regarding environmental impacts. Although DDT concentrations in the Black Sea have declined, significant quantities of organochlorinated compounds are still present. An estimated amount of 53,000 ty^{-1} of oil products are transported from the Danube River into the sea (Vespremeanu and Golumbeanu, 2018). To this number, 30,000 ty^{-1} wastewaters and 15,000 ty^{-1} effluents from industrial sources have to be added. However, it is difficult to get an overall reliable estimate of pollution by oil due to intensive transportation of crude oil by tankers. These tankers load crude oil mainly in Russian and Romanian terminals and carry it to the Mediterranean Sea through the Bosporus Strait. Oil transportation leaves its mark in the marine environment. Large quantities of sewage of ship origin also end up in the basin.

Wastewater discharge is not a minor problem: it is estimated that 517 million cubic meters per year of wastewaters are released into the basin (Vespremeanu and Golumbeanu, 2018) containing nutrients in organic and inorganic form, detergents and germs, some of them pathogens. These wastewaters insufficiently treated or even untreated, increase eutrophication levels and contribute to pollution due to toxic substances as well as to the microbial pollution of coastal waters. Wastewaters are not only connected with eutrophication but they also exercise pressure on the health of fisheries: "... *the use of the Black Sea and its tributaries for the disposal of wastes is free (un-priced) and so this ecosystem service is overused, imposing external costs on the commercial fisheries*" (BSC, 2008). However, the main reason of the collapse of the fish stocks in the Black Sea is overfishing. The large number of fishing vessels (about 4,000 fishing boats in 1997) and the use of crude techniques have caused a serious decline of fish populations with high commercial value (sturgeons and turbot). Landings amounting to 850,000 tons in 1985, declined to 250,000 in 1991; a partial recovery was recorded in 1995 (517,000 tons) mainly due to the anchovy population in the offshore waters of Turkey. The number of fish species with industrial interest was reduced from 26 species to 5 species (Vespremeanu and Golumbeanu, 2018). The poor condition of the Black Sea fish stocks is also connected with the presence and excessive growth of the alien species *Mnemiopsis leidyi*.

Mnemiopsis leidyi is a carnivorous planktonic species consuming zooplankton that includes crustaceans and various species of comb jellies but the main threat for the whole Black Sea ecosystem is the consumption of eggs and larvae of fish. This species was introduced in the Black Sea during the 1980s, probably through ballast water. It was first recorded in 1982. In spite of its predators (birds, fish, gelatinous zooplankton, scyphozoan), *Mnemiopsis leidyi* spread and reached its highest abundance in 1989: a population density of 400 specimens per m^3 of *Mnemiopsis leidyi* was recorded in places with optimal growth conditions. As *Mnemiopsis leidyi* consumes eggs and larvae of pelagic fish, its presence has contributed to the dramatic drop in fish populations of *Engraulis encrasicholus* (locally known as "namsi"). Biological control was attempted using another jelly fish species *Beroe ovata* but the success was not satisfactory. The species *Rapana thomasiana* is another alien species for the Black Sea environment originating from the Sea of Japan; it was soon well adapted and became competitor of many local mollusk species. Some of them disappeared.

There was exponential growth of the zooplankton biomass from 2.56 mgm^{-3} in 1961 to 8,719 mgm^{-3} (Bakan and Buyukgongur, 2000). This new regime favored an unusual dominance of some species such as *Noctiluca miliaris, Acartia clausi* and *Pleopsis polyphemoides*; nowadays 160 species are in danger.

4.3.3 Eutrophic conditions in the Black Sea

The Black Sea has been transformed over the last 40 years from a diverse ecosystem characterized by varied marine life into a eutrophic plankton based ecosystem (UNEP, 2006). According to the Global International Waters Assessment (GIWA), in the case of the Black Sea the most critical environmental issue, far more serious than any other environmental problem is eutrophication (UNEP, 2006). The first eutrophication phenomena in the Black Sea were observed during the 1960s (Bakan and Buyukgongur, 2000). The enrichment phase was followed by changes in the phytoplankton abundance and community structure, influencing in the end all trophic levels of the Black Sea ecosystem structure and function. Nowadays, the annual discharges of nutrients are roughly 160,000 tons of phosphorus and almost 800,000 tons of nitrogen. The main sources of phosphorus have been partitioned as follows: agriculture 15 per cent, wastewaters 46 per cent, industry 15 per cent and atmosphere 8 per cent. Nitrogen from agricultural sources 31 per cent, wastewater 26 per cent, industry 17 per cent and atmosphere 19 per cent. Enormous phytoplankton biomass growth reduces water transparency and the decomposition capacity exceeds the surplus deposit leading to anoxic conditions near the bottom. Under those conditions the benthic macrophytes disappear including the benthonic system of *Phyllophora* characteristic of the North-Western shelf that has almost been destroyed by now (Bakan and Buyukgongur, 2000).

Extensive eutrophic areas are located in the North-Western part of the Black Sea, the Azov Sea and the sections of Danube and Dnipro Rivers near the sea. A hypoxic area of 14,000 km^3 covering the shallow parts of the Black Sea was identified in 2000 (UNEP, 2006). As a result of excessive organic matter supply, phytoplankton blooms are usually dominated by toxic algal species, often occurring over the last 30 years. These eutrophic conditions had negative effects on species diversity. In addition, invasive species managed to get established in the basin, accentuating the impacts from eutrophication. Dam construction in the rivers reduced the quantities of silica outflowing in the sea, a situation that happens not only in the Black Sea but all over the world where river dams are built. Changes in the nutrient ratios N:P:Si where concentrations of nitrogen and phosphorus over silica concentrations are very high, trigger the formation of HABs with indirect effects, *inter alia*, hypoxia and fish death (Strokal et al., 2014). Even in cases where changes in the N:P:Si ratios are moderate in favor of N and P, there is a change in the diatom community structure where the dominance of relatively large diatoms, for example *Thalassiosira oestrupii*, is succeeded by relatively small diatoms like *Cyclotella choctawhatcheeana* (Mousing et al., 2015). Waters characterized by silica limitation, favor the growth of dinoflagellates, coccolithophores and small flaggellates (Eker et al., 1999). Deterioration of macrophytes formations in shallow waters affects the abundance of invertebrates and their diversity (Zaitsev, 1994). Furthermore, there is a shift towards smaller zooplankton groups, opportunistic and gelatinous species, less valuable for the higher trophic levels from the nutritive point of view (Kideys et al., 2000; Oguz,

2005). The control of mussels on phytoplankton biomass is poor as the result of the decline of the mussel population (Yunev et al., 2007). Nutrient inputs in the Black Sea during the period 1973 and 1990 have caused an estimated loss of living marine resources of about 60 million tons. Marine environmental degradation also has a negative impact on the revenue from tourism. The annual decline in the tourism revenue has been estimated to be 360 million $USD. Annual losses of 120 million $USD have also been estimated from reduced fishing yields. In spite of measures taken to mitigate the impact from eutrophication, both marine eutrophication and HABs are expected to increase in frequency and intensity over the 2020s.

Satellite surface chlorophyll distribution of the whole Black Sea area is shown in Figure 4.4. The Sea of Azov and the estuarine system of the Danube River are characterized as eutrophic areas. There are also some coastal eutrophic areas along the Turkish coasts as well as in the Sea of Marmara. The remaining area of the Black Sea is characterized as upper mesotrophic.

The structure of phytoplankton community has been variable between 2005 and 2010. Coccolithophores, especially *Emiliana huxleyi*, showed a dominance exceeding 90 per cent with biomass values ranging between 7.79 and 43.36 mgcm^{-3} (Silkin et al., 2014). On the contrary, the period 2007–2009 has been characterized by the dominance of diatoms, their biomass ranging between 48.81 mgcm^{-3} and 84.93 mgcm^{-3}. During this period of time, the diatom *Chaetoceros curvisetus* dominated over the shelf, whereas the diatoms *Pseudosolenia calcar-avis* and *Dactyliosolen fragilissimus* dominated in the open sea waters. Detailed information on phytoplankton succession along a transect from the Golubaya Bay to the center of the basin during 2005–2010 can be found in the work of Silkin et al. (2014).

Similar trends were obverved during 2002–2012 in the North-Eastern and Central parts of the Black Sea (Mikelyan et al., 2015). Satellite images were used to analyze coccolithophores blooms dynamics. Most of the blooms occurred in coastal

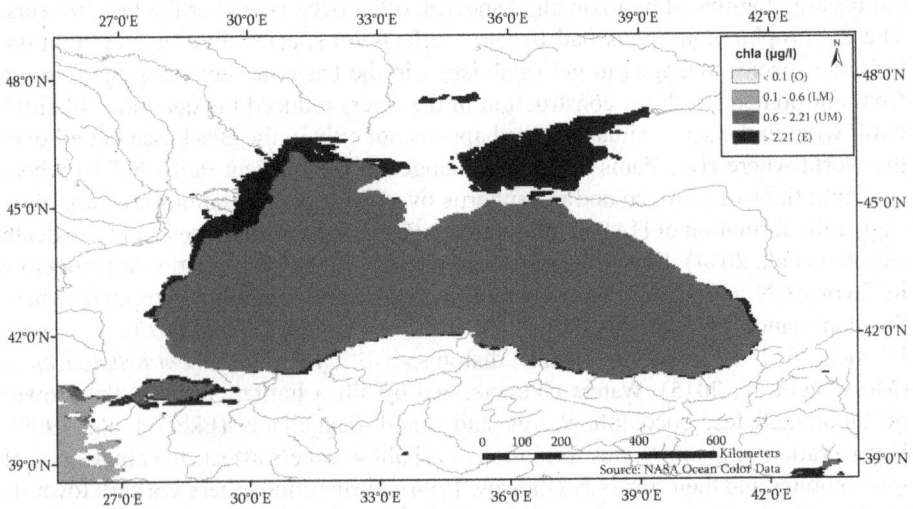

Figure 4.4 Satellite surface chlorophyll distribution in the Black Sea during 2017. Eutrophication scaling according to Simboura et al. (2005-modified) scale.

waters indicating the role of nutrient discharges from land-based sources. There are alterations in the dominance between diatoms and coccolithophores and it was found that the high diatom populations were the response of the system to high ammonium concentrations. On the other hand, high phosphorous concentrations favored the growth of coccolithophores. Thirteen phytoplankton species were contributing to the community structure during the period of study; the most dominant were: the coccolithophores *Emiliana huxleyi* and *Coccolithus* sp., the diatoms *Pseudonitschia delicatissima, Chaetoceros curvisetus, Chaetoceros affinis* and the dinoflagellates *Akashiwo sanguinea, Scrippsiella trochoidea, Gyrodinium spirale* and *Prorocentrum micans*. More recent information about nutrient, chlα and phytoplankton abundance in the Black Sea can be found in literature (Yalcun et al., 2017; Mikaelyan et al., 2017; Yunev et al., 2017).

4.3.4 Policy effectiveness in the Black Sea

The Black Sea unique characteristics are at risk due to extreme fisheries and pollution: views have been expressed that the sea was "dead", "one of the largest ecological catastrophes of our times", "close to collapse", etc. (Woodard, 1997). Following the collapse of USSR, a cluster of Treaties and Conventions were signed among the Black Sea riparian states. However, the problems are still remaining. There is an argument as to what is needed: more law or more policy? (DiMento and Hickman, 2012). Although cooperation at non-governmental level exists that is "citizen diplomacy" (DiMento, 1995), problems between states seem to be an obstacle in implementing measures of marine protection.

This section will be confined to assess the effectiveness of the Strategic Action Plan with emphasis on coastal eutrophication. Belarus, Russia and Ukraine have signed the Dnipro basin Environmental Programme in 1996. A number of memoranda have also been signed among the states regarding nutrients and eutrophication (Strokal and Kroeze, 2013). The results of many surveys indicate that nutrient discharges from rivers draining into the Northern part of the Black Sea and the Azov Sea have shown a decrease. This has been interpreted as the result of agricultural practices and compliance (to an extent) to environmental policy. However, this success will not be long lasting: sewage inputs increase annually due to coastal population growth and tourism: GDP per capita has been increased by 40 per cent and at the coastal cities has also increased especially from Asian regions (TDA, 2007; BSC, 2008).

Another potential source of nutrient of agricultural origin is from the bioenergy crop. There is an upward trend for this type of farming that requires fertilizers at least 10 per cent more than the traditional crops (Elbersen et al., 2009; FAO, 2005; Semerci et al., 2007). The Black Sea Strategic Action Plans (BS, SAPs) of 1996 and 2009 (BSC, 2008; BSC, 2009) include management options in the Black Sea Drainage Basin for agriculture and sewage. The whole idea is the reduction of nutrient inputs using as a baseline for nutrient concentrations the year 1960. This target requires "river specific policies". However, national policies that also include sewage and industry cannot be widely accepted to the same extent by all states draining into the river. The degree of collaboration is rather limited. Among the Black Sea rivers, the

Danube River has received considerable policy attention but reduction strategies still need to be implemented to satisfactory levels.

In conclusion, coastal eutrophication will be a problem in the future in spite of the efforts to mitigate nutrient discharges. There is an increase in dissolved N and P inputs in the Southern Black Sea due to an increase of sewage inputs. The effectiveness in reducing nutrients of land origin in the Black Sea may increase if "basin specific environmental policies" are adopted. Such a policy should take into account spatial characteristics of the basin as well as hydrology and human activities along the coastal zone. Models can also contribute towards a better understanding of the basin functions and in addition they can be useful management tools.

4.4 Eutrophication Status in the Caspian Sea

4.4.1 *Physiography of the Caspian Sea*

The Caspian Sea is the biggest landlocked endorheic sea of the world. It is located between the eastern part of Caucasus Mountain and the southeastern part of Europe. It is surrounded by Russia, Azerbaijan, Iran, Turkmenistan and Kazakhstan. The area of the water surface is about 400,000 km^2 and the water volume 80,000 km^3 whereas the length is 1200 km and the width ranges between 170 and 450 km. The total length of the shoreline is 7,000 km and the average depth 180 m. The maximum depth, 1,024 m, is located in a depression in the southern part of the Caspian Sea. The Caspian Sea morphologically is subdivided into three sub-basins due to its topography and not

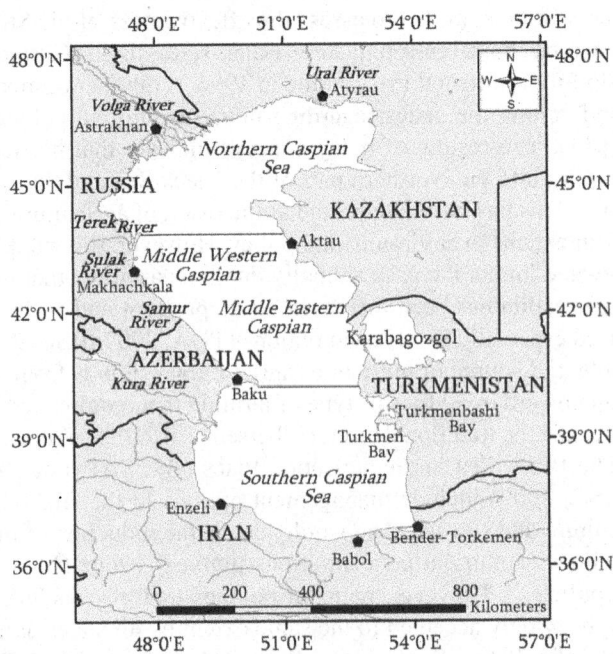

Figure 4.5 The Caspian Sea and the three sub-basins. (1) the North Caspian, (2) the Middle Caspian and (3) the South Caspian.

because of hydrodynamic reasons (Figure 4.5). The flat bottom of the northern part of the Caspian Sea is the shallowest part of the Sea. The middle basin is characterized by uneven bottom topography, whereas the southern part is dominated by the Caspian Depression. The proportional volumes of the three sub-basins, northern, middle and southern correspond to a total volume are 1/100, 1/3 and 2/3, respectively. The water level of the Caspian Sea is about 26 m below the sea surface area.

There are about 130 rivers inflowing into the Caspian Sea; most of them are small and only six rivers have a significant contribution to the water balance of the basin. The primary inflows into the Caspian Sea are from the rivers Volga, Ural, Terek, Sulak, Samur and Kura. There is also a connection between the Caspian Sea and the Azov Sea through the Volga-Don canal. The area of the Caspian catchment basin is about 3.5 million km²; this means that the ratio area catchment area to the sea surface area is 10:1 showing the significant role of the river water in influencing the hydrographic and environmental conditions of the Caspian system. The most important inflow is from the Volga River; Volga's River catchment area is 1.4 million km² which accounts for about 40 per cent of the total catchment area. Inflow distributions around the Caspian Sea are irregular, most of the fresh water originating from the northern and the western areas. Water inflows from Volga, Ural, Terek, Sulak and Kura is about 90 per cent of the total runoff into the Caspian.

The Caspian climatic conditions are rather complex as this sea extends over several climatic zones. Temperate weather conditions prevail in the northern part, whereas the climate in the western coasts is rather a moderately warm climate. The southwestern and southern regions are in a subtropical zone and the climate along the eastern coasts is characterized as a desert climate. More information on the physico-geographical conditions of the Caspian Sea have been given by Kosarev (2005).

The circulation pattern in the Caspian Sea is also complex. The general water circulation pattern in the deep-water areas of the Sea features a sub-basin eddy character. The dominant circulation pattern in the Middle Caspian is a dipole system: there is a cyclonic gyre in the northwestern area of the Sea and an anticyclonic gyre in its southeastern part. This dipolic circulation pattern dominates in the Southern Caspian Sea; however, the anticyclonic circulation extends over the northwestern part of the basin and the cyclonic circulation characterizes the southeastern part (Tuzhilkin and Kosarev, 2005). These cyclonic and anticyclonic water movements are the main reasons why oil pollution and eutrophication form transboundary environmental problems in the Caspian Sea.

4.4.2 *Environmental pressures and sea conditions of the Caspian Sea*

Urban centers around the Caspian shore vary significantly. Baku is the largest city and capital of Azerbaijan with a population exceeding 2,000,000. The economy of Baku is based on petroleum and petroleum exports. In addition, Baku is an important tourist destination in the areas of Caucasus. It is followed by Astrakhan, a Russian city built on the two banks of Volga River with a population exceeding half a million people. The city of Makhachkala, capital of the Republic of Daghestan, is having a population about 400,000. Along the Kazakh coast there are two regional centers

Aktau and Atyrau, each with a population of about 150,000. There are also cities along the Iranian coast, Enzeli (population 554,000), Babol (population 55,000) and Bender-Torkemen (population 173,000) (Zonn, 2005). Most of the population in Azerbaijan and Turkmenistan is concentrated within a 5 km wide coastal zone and the majority of the population is engaged in the oil industry and agriculture. Coastal activities as well as activities along the main Caspian rivers have a serious impact on the quality of the marine environment. The main sources of marine pollution are transboundary water pollutants, atmospheric deposition of toxic substances, discharge of industrial and agricultural wastewaters, disposal of domestic sewage and hydrocarbon compounds from drilling, extraction, transportation and refining. The daily production of crude oil in the area is about 1.5 billion barrels (Jafari, 2010). Once the spilled oil remains in the sea its toxicity usually increases due to photo-oxidation and biodegradation (chemical and bacterial): both processes increase the bioavailability of hydrocarbons; deposition of tarry pieces in the sediments also affects the benthic communities (Kitsiou and Karydis, 2014). Generally speaking, the Caspian marine environment is characterized as one of the most polluted among the world's regional seas. Ranking the pollution problems, oil pollution is first in the list, followed by eutrophication. Pesticides and heavy metals are not insignificant sources of pollution.

The Volga River (3,700 km long) flows through agricultural and industrial regions and therefore the waters discharged into the sea contain petroleum hydrocarbons, phenols, surfactants and organochlorinated pesticides; all these compounds threaten marine quality and ecosystem health in the Caspian. There are estimates regarding the release of chemicals into the marine environment: 60,000 metric tons of petroleum hydrocarbons, 24,000 tons of sulphites and 400,000 tons of chlorine end annually into the Caspian Sea (Aghai Diba, 2003). Concentrations of phenols, pesticides and heavy metals in the water column have been given by Korshenko and Gul (2005).

Serious threats come also from the construction of dam and reservoir systems: these flooded areas are not accessible to rare species such as beluga and sturgeon, famous for its caviar and the white fish belorybitsa: although they both live in the Caspian Sea, they spawn in the Volga River and its tributaries. Another economic activity which is under threat is the fishing industry. The Caspian caviar is considered as the most exquisite delicacy in the world. The price of the Caspian Beluga is about 1/5th of the average gold price. Catchments of sturgeon have decreased seriously over the last forty years. Landings from 30,000 tons in 1985, declined to 13,000 tons in 1990 and 2,100 in 1994 (Karydis, 2017). There are also species invasions (alien species) introduced by human intervention, usually through the river channel system that connects the Caspian Sea with the Azov and Black Sea, the Mediterranean and finally the World Ocean (Aladin and Plotnikov, 2004).

The jelly fish *Mnemiopsis leidyi*, an alien species of the Black Sea already mentioned in the Section 4.3.2, was transported to the Caspian Sea from the Black Sea, through the Volga-Don canal during the 1990s, most possibly in the ballast water cargo boats traversing the canal. Although the period of settlement is relatively short, this carnivorous jelly fish achieved a high density, exercising pressure on zooplankton decreasing its biomass. The decreased zooplankton biomass allows phytoplankton to thrive, observed as high chlα concentrations (Nezlin, 2005). This comb jelly has caused serious changes in the whole marine ecosystem of the Caspian Sea (Karpinsky et al., 2005).

4.4.3 Eutrophic conditions in the Caspian Sea

The most important impact on the Caspian Sea from the point of view of eutrophication, is the transportation of nutrients and biogenous substances through the river flow into the basin. It has been estimated that 40×10^3 tons of phosphorus and 400×10^3 tons of nitrogen per year on an average are getting discharged into the Caspian Sea. These discharges originate from agricultural runoff and the washout of irrigated farmland. The effects are particularly noticeable in the Northern part: more than 80 per cent of the area is affected by eutrophication of anthropogenic origin. Eutrophication symptoms are more pronounced during the summer where hypoxia in the bottom water layer occurs fairly often. The marine areas suffering from hypoxia have been estimated to be about $6-10 \times 10^3$ km² (Zonn, 2005). Eutrophic areas have also been reported in the eastern parts of the Caspian Sea where inflow of waste waters from the ports of Actau and Turkmenistan cause eutrophic conditions in the Kazkah, Turkmenbashi and Turkmen Bays.

Annual surface chlorophyll distributions based on satellite data is given in Figure 4.6. The Northern and Middle parts are characterized as eutrophic whereas most of the areas in the southern part is characterized as lower and upper mesotrophic. However, there are some coastal areas showing eutrophic trends.

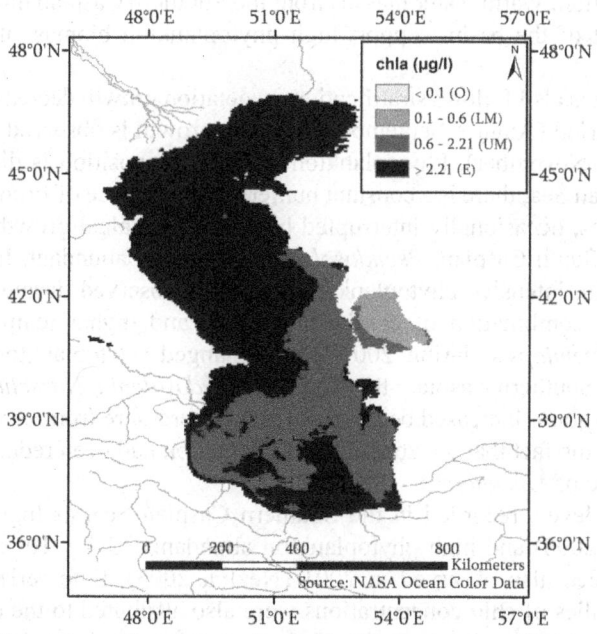

Figure 4.6 Satellite surface chlorophyll distribution in the Caspian Sea during 2017. Eutrophication scaling according to Simboura et al. (2005-modified) scale.

Phytoplankton distribution in the Caspian Sea is also influenced by the hydrographic regime of the basin. The Northern part of the Caspian Basin, receiving enormous amounts of river waters is dominated by estuarine freshwater species, whereas the predominance of euryhaline and brakish water species has been recorded in the Middle and Southern Caspian Sea. Some phytoplanktonic species are particularly abundant (Karpinsky et al., 2005), decreasing therefore, the diversity. A diatom usually showing high dominance is the species *Rhizosolenia calcar-avis*; as this diatom is not grazed by the marine zooplankton, tends to outgrow and is eventually deposited on the bottom of the sea where organic matter accumulates. Other species of phytoplankton showing dominance are the diatom *Actinocyclus ehrenbergii* and the pyrophyte *Exuviella cordata*.

There is spatial and seasonal distribution of phytoplankton in the Caspian Sea. During the summer, there is a massive growth of blue-green algae in the Northern Caspian, frequently caused by algal blooms. High values of phytoplankton biomass sometimes exceeding 10 gm^{-3}, have been measured near the Volga Delta (Karpinsky et al., 2005). Diatoms predominate in the Middle Caspian Sea during winter. The most abundant genus is *Pseudosolenia* accounting for 50 to 80 per cent in the phytoplankton biomass, whereas this dominance can reach 100 per cent of the biomass under special circumstances. During winter, the biomass of *Pseudosolenia* can be as high as two or three times compared to its summer abundance. The winter diatom regime of the Northern Caspian is succeeded by pyrophytes such as *Peridinium* and *Prorocentrum* during the spring period. Phytoplankton growth in the Middle Caspian is highly dependent on water stratification: there is intensive growth of microalgae during the mixing period, followed by a decline during the summer, due to stratification. Warm water masses from the Southern Caspian moving towards the eastern part of the basin, support high phytoplankton biomass in the Eastern Caspian Sea.

Phytobenthos also follows stratification: vegetation growth decreases during the stratification period (August–September) and new growth is observed when mixing starts (October–November). Phytoplankton species composition is different in the Southern Caspian Sea; there is a constant numerical dominance of *Prorocentrum* and other pyrophytes, occasionally interrupted by blue-green algal growth. Among the diatoms in the South Caspian, *Pseudosolenia* is the most abundant. In the western part of the basin, intensive phytoplankton growth is observed during the summer, favored by the combination of continental runoff and higher temperatures. The presence of *Mnemiopsis* during 2001–2003, changed phytoplankton community structure in the Southern Caspian: the diatoms *Pseudosolenia, Nitzschia seriata* and *Cerataulina bergonii*, increased because the grazing pressure from zooplankton was lessened due to the fact that the zooplankton population had been reduced due to the grazing pressure of *Mnemiopsis*.

High chlα levels recorded in the Southern Caspian Sea, as high as 9 mgm^{-3} (Kideys et al., 2008) and high phytoplankton abundance, 3.4×10^3 cellsl^{-1}. These maxima had been also observed in 2001 (Nezlin, 2005); time series analysis of seasonal anomalies of chlα concentrations were also attributed to the consequences of the jelly fish *Mnemiopsis* as mentioned above.

4.4.4 Policy effectiveness in the Caspian Sea

The main environmental tool for environmental protection and management of the Caspian Sea is the Caspian Environmental Program (CEP). CEP has established a number of centers dealing with the transboundary pollution problems in the area. The development of transboundary water management projects was aiming at improving the capacity of the states involved in collective management of the shared resources. In addition, to the effects of International Organizations, the Caspian States should also draft National Action Plans and implement them parallel to the Strategic Action Plans. Beyond any international initiatives, political and economic conflicts among the riparian states should be taken into account. The Caspian Sea is the third oil field in the World and the intensive exploitation of oil by Azerbaijan, Turkmenistan and Kazakhstan, threatens the environment with serious oil pollution problems. The overall impact on the ecosystem is a combination of eutrophic conditions and oil pollution; it is therefore necessary that in any Action Plan, these two problems have to be taken into account concurrently. Another problem is the economic consequences of reducing nutrient discharges into the marine environment as it differs among the Caspian States: more than 85 per cent of the river inflow into the Caspian originates from Russia. On the contrary, Iran's contribution to the Caspian Sea in terms of nutrients and eutrophication is negligible.

As a conclusion, referring to the effectiveness of the measures, it has to be said that there are not any significant improvements in the quality of the environment. The unsettled allocation of responsibilities and the uncertainty regarding the riparian states of the Caspian Sea are the main obstacles for any measures of protection and management of the marine environment to be successful. The main activities in the Caspian region are oil production, agriculture, fishing and tourism; the priorities of these activities are not proportionally equal to all States, which is an additional impediment in finding a balanced solution combining socio-economic interests and nature-conservation interests that would satisfy all the Caspian States. Kostianoy and Kosarev (2005) have already proposed that the key issue should be *"the elaboration of a system of coordinated state and interstate mechanisms, actions and guarantees based on the compliance, by one and all states, with the common humanitarian principles and norms of international legislation that are called to guarantee effective solutions or to prevent emergence of environmental problems of interstate and world community dimensions"*. Even so the problem of oil pollution is a priority for the Caspian environment. Furthermore, mitigating eutrophication is an issue with high economic and even higher social cost. Therefore, the effectiveness of the Caspian Environmental Policy will be a long awaiting objective.

4.5 Eutrophication Status of the Baltic Sea

4.5.1 Physiography of the Baltic Sea

The Baltic Sea is a European landlocked sea, one of the largest brackish water systems in the world (Karydis and Kitsiou, 2014). The surface area is about 350,000 km^2, the volume being about 20,000 km^3. The coastline length is about 8,000 km and the average depth is about 50 m but in more than 50 per cent of the area, it is less than that. The maximum depth, 459 m is in the center of the basin towards the Swedish side. The catchment area is about 1,600,000 km². The Baltic is divided into five sub-basins (Figure 4.7): (a) the Baltic Proper (211,069 km²) (b) the Gulf of Bothnia (115,516 km²) (c) the Gulf of Finland (29,600 km²) (d) the Gulf of Riga (16,300 km²) and (e) the Danish Straits (42,408 km²). There is also the transition zone to the Skagerrak–North Sea region. The basin is surrounded by 14 countries including Denmark and Sweden in the west, Finland in the north, Estonia, Russia, Latvia and Lithuania to the southeast and Germany, Poland to the south (DiMento and Hickman, 2012).

The hydrology of the basin is influenced greatly by freshwater inflows. There are more than 250 streams draining the catchment area, contributing about 660 km³ of fresh water per year. The major rivers in the area are the Oder River, the Vistula River, the Neman River, the Daugava River and the Neva River. The precipitation–evaporation budget is positive. Due to the brackish water supply, there is a continuous flow out to the North Sea through the Danish Straits to Kattegat and eventually to the Atlantic. Salt water from the North Sea inflows in the Baltic. The ecological significance of the inflowing waters is noteworthy because these oxygen rich waters support the benthic communities in the Baltic deeps. Salinity values of the outflowing water range between 3.0 and 9.0‰, whereas the salinity of the inflowing North Sea water is 35 psu. This difference in salinity between surface and bottom waters creates vertical stratification. The halocline forms a barrier for nutrient and oxygen exchange.

4.5.2 *Environmental pressures and sea conditions in the Baltic Sea*

Pressures in the Baltic Sea originate from three different sources: (a) nutrient input from the rivers and the catchment area of the Basin as well as domestic sewage (b) chemical pollution and (c) overfishing (DiMento and Hickman, 2012; Karydis, 2017).

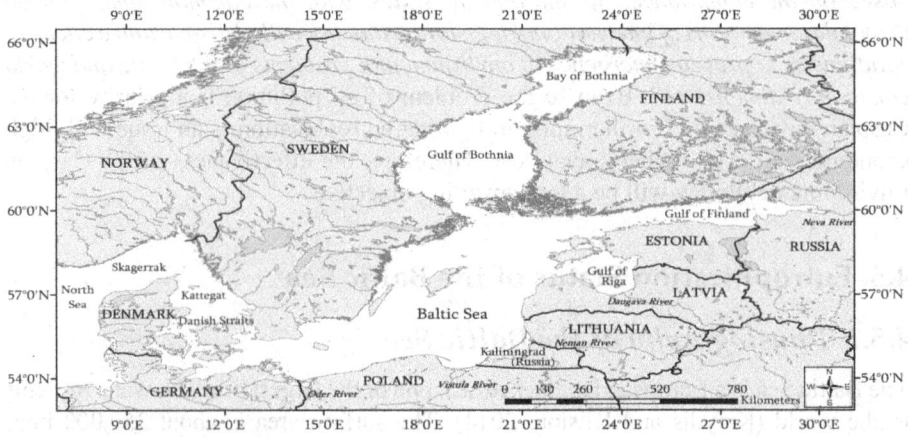

Figure 4.7 The Baltic Sea and its sub-basins.

The Baltic Sea receives industrial effluents containing PCB compounds, pesticides (DDT, HCB and lindane) as well as TBTs and various heavy metals including lead. Declining trends have been recorded over the last few years. However, there are emerging pollutants with close links with the modern lifestyle. An additional environmental problem in the Baltic is high concentrations of cesium-137 from the Chernobyl accident in 1984 as well as relatively high concentrations of strontium-90 from atmospheric tests of nuclear weapons during the 1950s and 1960s (DiMento and Hickman, 2012).

Fish stock including herring and sprat are suffering from overfishing (Aps and Lassen, 2010). At the same time the number of alien species is increasing and more than 100 species have been recorded. As these alien species have adapted to the special conditions of the Baltic Sea, species and habitats belonging to the native fauna and flora have become vulnerable (Johanhesson and Andre, 2006). Nutrient inputs and coastal eutrophication have been recognized as the main environmental problem in the Baltic Sea (Lundberg et al., 2005; Andersen et al., 2010; Flemming-Lehtinen, 2007). Excessive phytoplankton growth attenuates light penetration, making light unavailable in the deeper part of the water column. As a result, the submerged aquatic vegetation along the coastal zones dies (Chislock et al., 2013). The phenomenon of mass algal growth also affects recreational and tourism activities (Backer et al., 2010). In addition, algal toxins enhance health hazards to the public (Hunter, 1998).

The Baltic Sea area is inhabited by 85 million people. Population distribution is not homogeneous, ranging from 1 inhabitant per km^2 in the northern and north-eastern part to 100 inhabitants per km^2 in the southern and south eastern parts. There is a variety of land-based human activities, the most important from the environmental point of view being agriculture, industrial and urban activities (HELCOM, 2009). Shipping activities among the surrounding countries, fish farming, gravel extraction and wind farms have also to be taken into account. The limited water exchange between the Baltic and the North Sea has intensified some of these pressures, including nutrient drainage and their accumulation in both organic and inorganic forms. HELCOM (2017) has identified seven principal pressures: (a) eutrophication (b) hazardous substances (c) marine litter (d) underwater sound (e) non-indigenous species (f) species removal by fishing and hunting (g) seabed loss and disturbance.

The principal environmental problem in the Baltic Sea is eutrophication. According to the Integrated Status Assessment (HELCOM, 2009), 97 per cent of the region has been characterized as eutrophic. Although nutrient inputs have been reduced as a result of measures taken, the effectiveness of these measures has not been documented so far by the ecological surveys carried out in the area. Nitrogen and phosphorus concentrations showed an increase up to the late 1980s; this increase was attributed to increased nutrient loading from the terrestrial environment since the 1950s (Gustafsson et al., 2012). Due to improved waste water treatment and suitable agricultural practices, reduced nutrient loading was recorded in the 1990s. Total nitrogen and total phosphorus loads have decreased by 16 and 18 per cent respectively, between 1994 and 2010; it is also possible that these changes may be greater in individual sub-basins (HELCOM, 2013). However, the significance of nutrient run-off from cultivated land has not been identified (HELCOM, 1996) as a highly noteworthy nutrient source in the Baltic Sea. Since the 1990s, reductions have been observed in nutrients and strong reduction has been recorded recently. However, in spite of the progress, time series since the 1990s in the Baltic Proper have shown that deteriorating symptoms over the early 1990s were still present.

Hazardous substances that enter the Baltic Sea are mainly man-made chemicals and heavy metals; they are leaching from domestic materials, through waste water treatment plants, waste deposits and atmospheric deposition of emissions originating from industrial sources. There are many different types of impact on the marine organisms. With the exception of oil, which is visible as oil-spills most of these toxic compounds cannot be seen in the marine environment. Most of them degrade very slowly and accumulate in sediments and the aquatic food web. Among the principal hazardous chemicals in the area are mercury and polybrominated flame retardants. Other hazardous substances in the Baltic that have received the attention of marine scientists are hexabromocyclododecane, polybrominated diphenyl ethers, PCBs, dioxin, furans, polyaromatic hydrocarbons and the products of biodegradation, as well as a number of heavy metals. Cadmium and mercury also enter the Baltic Sea through atmospheric deposition. There are also emerging pollutants which are pharmaceuticals entering the Baltic Sea either biologically (via urine and/or feces) or through inappropriate disposal.

The visible marine litter in the Baltic is a problem as it appears on the beaches, the shoreline, the water column and the bottom of the sea. Artificial polymers, known by the public as "plastics", form 70 per cent of the marine litter. Litter in the sea is having socio-economic impact as well as impact on the marine organisms. It can be toxic and sometimes it is ingested by animals. It can also damage marine habitats. Plastics degrade slowly and the microscopic segments are ingested at the bottom of the food pyramid causing even more serious problems. Marine litter also causes problems on fishing gear and the navigation of small boats.

Although the underwater sound is produced naturally by wind, waves, ice and thunders, it can also be produced by humans through wind sea farms (Karydis, 2013), underwater constructions, sonars and seismic guns just to mention a few. Due to the impact on marine life (Carstensen et al., 2006), the Baltic States have agreed to register these activities in order to be under control. This information helps evaluation regarding possible impacts on marine life, especially marine mammals.

Various non-indigenous species have settled in the Black Sea environment as a result, either of intentional or unintentional human activities (shipping, hazards from the ballast water) or aquaculture; they are both the major routes of species invasions. So far 140 non-indigenous species have been registered in the Baltic and of these, 14 have been recorded during 2011–2015. The main impact from alien species is potential changes in the structure and function of the marine ecosystem, biofouling of boats and intake-outflow pipelines, public health impacts, if pathogens are involved, fishing problems, and finally economic impact.

Species removal by fishing and hunting, although traditional sources of livelihood in all the states surrounding the Baltic Sea, may lead to a reduction or even extinction of fish stocks in the area. Seabird loss and disturbance or even loss of biodiversity are additional threats in the Baltic environment. All these forms of pressure should be taken into account when eutrophication problems are addressed, as in many cases the effects are synergetic.

4.5.3 *Eutrophic conditions in the Baltic Sea*

Eutrophication in the Baltic Sea has been recognized as a serious problem since the mid-1990s including several negative symptoms regarding ecosystem quality (Larson et al., 1985; Bonsdorff et al., 1997). The main objective of the Baltic Sea Action Plan is to reduce eutrophication in the Baltic to a great extent. As a result of this Action Plan, many eutrophication assessments have been performed (HELCOM, 2009; HELCOM, 2010; HELCOM, 2014). The assessments were carried out using three indicators: nutrient levels (DIN and DIP), direct effects (chlα concentrations and Secchi disk) and indirect effects (dissolved oxygen concentrations) (HELCOM, 2017). At the same time, threshold values were adopted so that the evaluation of any progress towards the targets set in the Action Plan to be evaluated. The nutrient status during 2001–2006 was characterized by DIN concentrations in the Bothnian Bay, a P-limited area. High DIN concentrations were also measured in the Gulf of Finland due to the Neva River discharges. Winter DIN concentrations in the other basins ranged between 3 and 4 μmol.l^{-1}. High DIP concentrations were measured in the Gulf of Riga and the Gulf of Finland, the values being 0.78 and 0.84 μmol.l^{-1} respectively. DIP concentrations during this period in the Bothnian Sea, the Baltic Proper and the Danish Straits, ranges from 0.36 to 0.47 μmol.l^{-1}. Further eutrophication assessments, referring to the period 2011–2015 showed that the Baltic Sea was still affected by eutrophic conditions (HELCOM, 2017). A total area equal to 97 per cent of the Baltic Sea was eutrophied and only some coastal areas had achieved good status. Nutrient levels were far from the threshold values, characterizing a good status, a fact that weighted on the results of the integrated assessment procedure (HELCOM, 2006). However, using the scale developed for the Mediterranean Sea (Simboura et al., 2005), it is obvious (Figure 4.8) that the surface waters can be considered as eutrophic in spite of the improvements in the water quality that have been recorded.

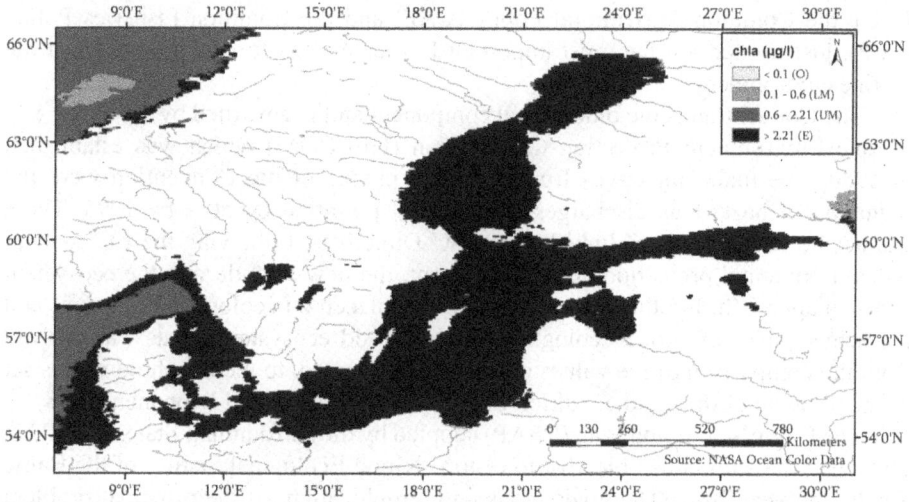

Figure 4.8 Satellite surface chlorophyll distribution in the Baltic Sea during 2017. Eutrophication scaling according to Simboura et al. (2005-modified) scale.

Phytoplankton easily responds to changes in nutrient levels and therefore, chlα concentrations is a suitable parameter for assessing phytoplankton levels. Phytoplankton biomass is a combination of the hydrological regime and eutrophication processes (Wasmund and Uhlig, 2003). Reference conditions for chlα values have been set for the summer time. The results during the period 2001–2006 showed substantial deviations in the Gulf of Finland, the Gulf of Riga and the Northern Baltic Proper from reference conditions values. Later, during 2011–2015, chlα and Secchi depth showed that they did not achieve threshold conditions in any open sub-basin east of the Baltic Sound. However, threshold values were achieved in the Kattegat and west of the Sound. In the Kattegat and the Danish Straits, chlα concentrations have been decreasing since the late 1980. Oxygen values did not approach threshold values in any open sub-basin.

As a conclusion, nutrient inputs have decreased steadily in most sub-basins of the Baltic since 1980s. On an average, this reduction is estimated between 16 and 18 per cent (HELCOM, 2014). Despite the reduced nutrient inputs, chlα trends still show no sign of decline. However, in some cases, an increase has been observed.

4.5.4 *Policy effectiveness in the Baltic Sea*

Marine environmental protection of the Baltic Sea is based on two legal components: (a) *"the Convention on the "Protection of the Marine Environment of the Baltic Sea Area"* enforced in 1974 and revised in 1992. This Convention has been signed by all surrounding States of the Baltic Sea Basin and (b) the EU Directives relevant to the marine environment. These days all the surrounding States, except Russia, are Member States of the EU and therefore, most of the Baltic catchment area is within the jurisdiction of the European Union. These States have to comply with a number of EU Directives such as the Framework Water Directive (Directive 200/60/EC), the Marine Strategy Directive (Directive 2008/56/EC), the Nitrates Directive (91/676/EEC), the Common Agricultural Policy (CAP) and the Common Fisheries Policy (CFP), just to mention the most important EU legal tools for the protection of the marine environment.

The Convention is the oldest legal component and is governed by the Baltic Sea Marine Environment Protection Commission (HELCOM) which was established in 1980. The main objectives from the early phases of the Convention were the reduction of hazardous discharges and nutrient pollution by 50% by 1995. There has been a restart in 2003 linked to the EU Directives. Following the EU concept on environmental protection and management, the new attitude was the ecosystem oriented approach. In other words, priority was placed on ecological objectives and the achievement of "good ecological status". Good ecosystem status was derived from the comparison of the values of the indicators used, to the threshold values set as the baseline of the good condition of the system (Backer and Leppanen, 2008).

The Baltic Sea Action Plan (BSAP) adopted by the surrounding States and EU in 2007, had the ambitious objective to restore "Good Ecological Status" of the Baltic Sea by the year 2021. The goals focus on eutrophication, as the principal problem in the Baltic, biodiversity, hazardous substances and maritime activities (HELCOM, 2017a). Regarding eutrophication, BSAP in order to reach good environmental health has adopted five ecological objectives to describe the ideal conditions of the Baltic Sea unaffected by eutrophication: (a) nutrient concentrations close to threshold values (b) clear water (c) natural levels of algal growth and (d) natural occurrence of plants and animals as well as their distribution. If the eutrophication objective fails to be achieved, the status of biodiversity will be also problematic. In addition, failure to reduce airborne nitrogen emissions from shipping, untreated sewage and nutrient emissions from ships will eliminate the success of the eutrophication objective. A nutrient reduction scheme was introduced in 2007 in the HELCOM Baltic Sea Action Plan. The riparian countries, have agreed on nutrient reduction targets between 2008 and 2013. There were two main components to implement the reduction scheme (a) Maximum Allowable Inputs: it refers to the maximal level of inputs of water and airborne nutrients (nitrogen, phosphorus) that will be allowed to enter the Baltic Sea and (b) Country Allocated Reduction Targets: indicates the quantity of nutrients that each riparian state should reduce, in order to achieve the objective. Although figures regarding nutrient input reduction vary (Karydis, 2017; HELCOM, 2017a; HELCOM, 2017b), there is a general agreement that the overall nutrient reduction is around 20 per cent. Part of this reduction is due to the improvements in wastewater treatment all over the Basin (DiMento and Hickman, 2012).

The excellent communication among Member States in the Baltic, their governments, the policy makers and the scientific community, indicates that the effectiveness due to good collaboration is high. However, during the early stages of the HELCOM implementation, scientific and technological cooperation was not always easy mainly due to the *"western"* and *"eastern"* political systems. Russia was very restrictive in sharing data with the neighboring countries and still is to some extent (DiMento and Hickman, 2012). Good communication among Member States in the Baltic, is also obvious from the fact that there are NGOs that have been enabled to participate as observers since 1992 in the HELCOM work.

The costs and benefits from reducing eutrophication in the Baltic Sea have been dealt with in a previous HELCOM Report (HELCOM, 2009). Any uniform reduction in nutrient load percentages among the riparian Baltic States has been considered as economically inefficient (Gren et al., 1997). It has been estimated that total costs for mitigating eutrophication with cost-effective solutions would range between €1,600 and €16,500 million per year (Turner et al., 1999; Gren et al., 1977; Wulff et al., 2001; HELCOM and NEFCO, 2007). On the other hand, the benefits referring to goods and services either with the use or no-use values, that are provided by the marine ecosystem are nutrient recycling, climate regulation, seafood production and recreational opportunities. Provision of those goods and services can be improved by reducing eutrophication. The Baltic Sea is a productive ecosystem and it has been estimated that ecosystem services provided by the marine environment worth between $2,000 and $3,000 per hectare (Costanza, 1997). The nutrient assimilative capacity in the Baltic is free of charge and extensively overused (Turner et al., 1999). Furthermore, it has been estimated that there would have been net benefits as a result of reducing eutrophication. These net benefits range from €48 million to €4,070 million per year (HELCOM and NEFCO, 2007).

In conclusion, it is commonly accepted by now that the Baltic cluster is the most complete and well developed among similar clusters in other regional seas.

A significant part of the Baltic Sea cluster is focusing on marine eutrophication, which is the most important problem in the areas. Eutrophication issues and their effectiveness have been divided into three components: the technical and the scientific components as well as the action-oriented component. The Baltic Sea countries share a common understanding about causes and effects of eutrophication; this has helped them to develop common methodological tools (indicators, parameters, monitoring schemes) as well as widely accepted assessment procedures. This allowed for most of the parameters to form long and reliable data time series.

The action-oriented issues are closely linked with the reduction of nutrient loads. Nutrient inputs have been reduced, as mentioned before; however, this reduction is not reflected at the eutrophication status of the area as yet. So the targets regarding mitigation of eutrophic conditions have not been reached so far. As changes in new agricultural practices are needed, the whole effort will take time and needs patience. At the same time, the implementation of the UWWTD, ND, WFD and MSFD Directives are an additional weaponry for improving the quality of the marine environment.

4.6 Eutrophication Status of the North Sea

4.6.1 Physiography of the North Sea

The North Sea is one of the marginal seas of the Atlantic Ocean surrounded by Denmark, Germany, the Netherlands, Belgium and France in the south. It is also surrounded by England and Scotland in the west and Norway in the east. It communicates with the Atlantic in the south through the English Channel and through the Norwegian Sea in the north. In the east, there is connection to the Baltic Sea through the Skagerrak and Kattegat, the narrow Straits that separate Denmark from Norway and Sweden respectively (Figure 4.9). The length of the North Sea is

Figure 4.9 The North Sea and NE Atlantic (OSPAR Area) and its regions.

more than 970 km and the width is 580 km. The area of the North Sea is 570,000 km^2. The North Sea volume is about 54,000 km^3 and the maximum depth is 700 m. This depth is located in the Norwegian Trench whereas the average depth of the North Sea above the European continental shelf is about 90 m.

Work on the physical oceanography of the North Sea and especially on the dynamics of water masses, currents and water balance has been published by Hill (1973), Lee (1980), Otto (1983) and Sundermann (1994). In the present chapter, a brief description of the hydrology regime will be given for a better understanding of eutrophication processes. Lee (1980) has proposed a classification scheme comprised of five water bodies: the North Atlantic, the Continental coast, the English coastal area, the Scottish coastal area and the Skagerrak. These water masses are delineated as "fronts" (sharp differences between water masses) which are easily distinguishable by satellite imagery. Salinity values range between 3 and 35 psu and the temperature varies between 6 and 17°C (Reid et al., 1988). Salinity values fluctuate in areas where there is inflow of river water. The major outflow into the North Sea comes from the Elbe and the Rhine-Meuse watershed.

The general circulation pattern is in anticlockwise rotation. The North Sea gets the majority of an ocean current from the northwest opening and a small water mass through the English Channel. Most of the surface waters characterized by low salinities move offshore whereas deeper water masses, more saline, move towards the shore. Freshwater inflows have been estimated to be as high as 5360 m^{3}s^{-1} (Taylor et al., 1981). British rivers peak their outflows in winter and minimize them during summer. Rivers flowing into the German Bight show maximum and minimum flow rates during spring and autumn respectively. The Rhine shows an intermediate pattern. Precipitation in the areas exceeds evaporation, the difference being about 2,700 m^{3}s^{-1}. This excess water increases from 70 mm in the southern part of the North Sea to 200–300 mm in the northern areas (Becker, 1981).

The marine environment in the North Sea is rich in zooplankton, dominated by copepods. Various crustaceans are also common in the area. There is commercial fishing of Norway lobster, brown shrimp and prawns. More than 230 species of fish have been identified, some of them commercially exploited like cod, haddock, whiting, sole, plaice, mackerel, herring, sprat and sand eel. Coastal waters, marine installations or small islands offer habitats to marine mammals like the common seals and harbor porpoises. In the Northern part, a variety of pinnipeds are present (harp, hooded and ringed seals) along with walruses. The cetaceans include porpoises, dolphins and different species of whales. There are many habitats in places like the Ythan Estuary, the Farne Islands in the UK and the national parks in the Wadden Sea suitable for seabirds. Atlantic puffins, northern fulmars, species of petrels, sea ducts and gannets, cormorants, gulls and terns being the most common species.

4.6.2 *Environmental pressures and sea conditions in the North Sea*

The North Sea is surrounded by a population of about 40 million people and heavily industrialized countries exercising a number of strong environmental pressures.

Population pressures increase during the summer months as tourists are added mainly in coastal areas. As a result, the North Sea is used intensively for a variety of activities. There is heavy maritime traffic in the Straits of Dover and the southern areas of the North Sea. Sand and gravel needed by the construction industry, is dredged in many places; natural gas and oil are extracted in the Northern and Central parts of the North Sea. There are oil inputs from offshore installations, from ships as well as accidental losses. Oil spills, in addition to biological and ecosystemic effects (Spies, 2011), they have a high public profile. The North Sea has not so far had major tanker accidents but some "blowouts" from offshore platforms did happen (Dicks et al., 1988). An additional pressure comes from the sewage, treated or untreated, from coastal areas and the rivers, mainly the Rhine, Elbe, Weser, Scheldt, Ems, Thames, Trent, Tees and Tyne rivers. Among the impact from industrial activities, heavy metal pollution is also a serious threat in the marine environment. The highly productive North Sea fisheries are also overexploited.

All these activities mentioned in the previous paragraph, pollute the North Sea environment via their inputs. Inputs of heavy metals can originate either from point-sources (dumping areas or estuaries) or non-point sources (atmosphere). Polluted sediments can also form a source of heavy metals, especially metals characterized by nutrient mobilization (Kersten et al., 1988). There is also an increased concern for atmospheric deposition. The industrialized states around the sea form considerable sources of atmospheric pollution, rich in metallic and metalloid compounds. Processes including dry deposition, wet deposition and chemical transboundary pollution from metallic compounds have been studied a long time ago (Kersten et al., 1988). There are more than 60,000 organic compounds in use. Potential sources of organic micropollutants ending in the marine environment are rivers, the atmosphere and shipping activities. These compounds, most of them highly toxic, are distributed among different "compartments", that is, water, organisms, sediments and suspended matter. Prominent is the role of hydrocarbons, polychlorinated biphenyls (PCBs) and hexachlorinates (HCHs).

However, the most serious pressure comes from nutrient inputs. Nitrogen discharges in the Greater North Sea are estimated between 1,400 and 2,000 kilotons per year. Out of this, a quantity of about 500 kilotons, that is, 24–28 per cent of the total amount, ends into the marine environment via the atmosphere (OSPAR, 2007). Some flood events in the European Continent enhanced nitrogen discharges in 1994, 1995 and 2002. Total nitrogen inputs in the Greater North Sea have been stabilized over the last 15 years, estimated about 1,500 kilotons per year. Phosphorus inputs also showed a downward trend since 2002. Before that year, the phosphorus inputs were ranging between 70 and 90 kilotons per year. Phosphorus inputs after the year 2002 were stabilized to 40 kilotons per year. Nutrient inputs from the Bay of Biscay and the Iberian Coast also showed a decline, nitrogen inputs did not exceed 300 kilotons per year since the year 2000 with a minimum of 170 kilotons in 2013. In the same areas, phosphorus inputs since 2004 were estimated to be 10 kilotons per year.

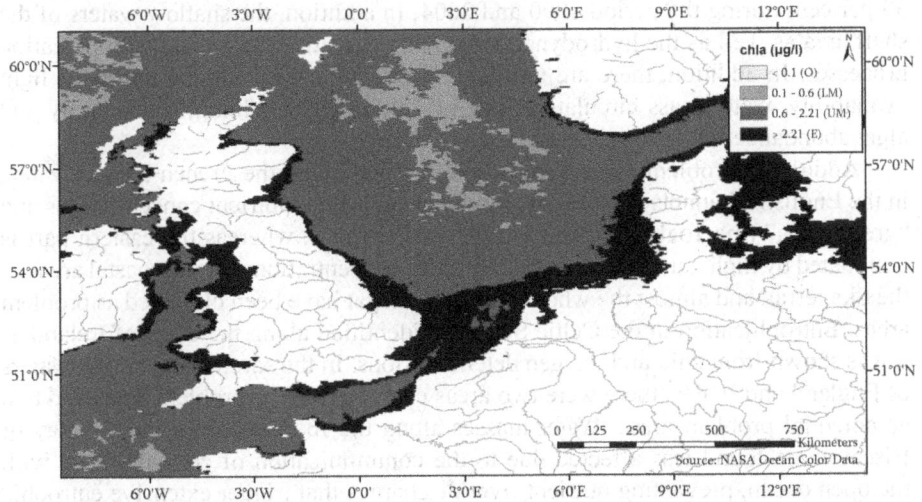

Figure 4.10 Satellite surface chlorophyll distribution in the North Sea during 2017. Eutrophication scaling according to Simboura et al. (2005-modified) scale.

4.6.3 Eutrophic conditions in the North Sea

The principal sources of nutrients in the OSPAR area that is the North Sea and the NE Atlantic are river outflows. The OSPAR Contracting Parties have applied a common procedure using data sources from 2006 to 2014 for assessing eutrophication levels (OSPAR, 2017). It was found out that eutrophication is still a problem (Figure 4.10). Seven per cent of the North Sea Atlantic Area has been characterized as eutrophic, mainly in coastal waters. The most outspread problem of eutrophication had been located in the Greater North Sea where 98,000 km² was identified as eutrophic, covering coastal waters from Belgium, the Netherlands, Germany, Denmark and Sweden. Coastal waters in France, Norway and the United Kingdom seemed to have minor eutrophication problems, covering an area about 400 km², whereas small areas in the Celtic Seas and the Bay of Biscay were eutrophic being about 500 and 800 km² respectively. For assessment purposes, the area was divided into five regions: Region I: Arctic, Region II: the Greater North Sea, Region III: the Celtic Seas, Region IV: the Bay of Biscay and Region V: the Wider Atlantic (the NE Atlantic area). The waters were classified as problem areas, potential problem areas and non-problem areas. The Norwegian Sea and the Barents Sea has been assessed as a non-problem area. On the contrary, in the Regions II, III and IV, many assessment areas have been characterized as eutrophic (problem areas). Most of them were in coastal areas and inshore waters particularly in fjords and estuaries. Overall 100,000 km² were classified as problem areas and about 25,000 km² as potential problem areas.

In the Greater North Sea, eutrophic conditions have been identified as well. This area is about 750,000 km² and almost 14 per cent of the area has been characterized as problem area and 3 per cent as potential problem area. Contribution from atmospheric nitrogen in the Greater North Sea (Region II) range between 25 and

39 per cent during the period 1990 and 2004. In addition, the shallow waters of the shelf area as well as the hydrodynamic conditions tend to accelerate eutrophication processes. In addition, there are multiple synergistic effects among nutrients, light availability, water mass circulation, mixing in the water column, productivity and algal abundance (Ducrotoy, 1999).

Additional problem areas have been identified along the French coastal waters in the English Channels (Figure 4.10); the reason is high nutrient concentrations and "green tides" (macroalgal blooms) in the western part, whereas the eastern part is dominated by high nutrient loads and high chlα concentrations. Many coastal areas of the Skagerrak and almost the whole area of Kattegat have been classified as problem areas. Eutrophication in the Celtic Seas was identified along the coast of Ireland as it was shown from chlα and oxygen determinations. In the south-west coastal waters of England and Wales, there were two areas characterized as problem areas and two as potential problem areas. Water masses along the Iberian coasts and the Bay of Biscay were found less affected due to the communication of coastal waters with the open ocean, preventing nutrient river discharges that trigger extensive eutrophic phenomena. Frequent phytoplankton blooms are an alarming symptom in the North Sea. The algal bloom episodes per year have been doubled over the past three decades. The severity of the eutrophic symptoms has also to be mentioned: oxygen depletion of water masses near the bottom is a common symptom.

The ample availability of nitrogen and phosphorus has led to silicon deficiency; silicon was abundant in the North Sea 40 years ago (Van Bennekom et al., 1975). Phosphorus reduction observed over the last few years in the Greater North Sea, has led to an overall change in the molar ratios between nitrogen and phosphorus. This ratio was 45:1 in 1990 and became 80:1 in 2014 (OSPAR, 2017). Similar changes have been observed in the Iberian Coast and the Bay of Biscay where the N:P ratio from 40:1 changed to 80:1, becoming the limiting nutrient (Lampert et al., 2002). The benthic-pelagic coupling in the North Sea is important for the ecosystem functioning due to the extensive shallow areas. Phytoplankton in the North Sea occurs up to a depth of 30 m, once the waters are transparent whereas their maximum depth of occurrence does not exceed depths between 5 and 10 m if water transparency is reduced. Accumulation of macrophytes has been observed over the last decades. It has been reported that 50,000 tons of plant material is removed every year from the North Sea beaches (NSTF, 1994). Domination of *Chlorophytes* and *Ulvae* has been observed. In the German coasts at least 15 per cent of the shallow areas are covered every year; sometimes the weed carpet can cover as much as 30 to 60 per cent of the tidal flats (Ducrotoy, 1999).

4.6.4 Policy effectiveness in the North Sea

The OSPAR Convention is the main legal and management tool for the States surrounding the North Sea. As a result of the experience gained about setting targets, collaboration among the Member States, assessment procedures and effectiveness of the measures, OSPAR has launched a strategy for the "*Protection of the Marine Environment of the North Sea–Northeast Atlantic 2010–2020*". The OSPAR Commission's concept was to harmonize its policy with the EU Directives (WFD,

MSFD, ND, UWWTD and the IPPC Directive) by implementing the Ecosystem Approach. According to OSPARs objectives *"this is a comprehensive integrated management of human activities based on the best available scientific knowledge about the ecosystem and its dynamics, in order to identify and take action on influences which are critical to the health of marine ecosystems, thereby achieving sustainable use of ecosystem goods and services and maintenance of ecosystem integrity"* (PART I: Implementing Ecosystem Approach in OSPAR Agreement, 2010). From the point of view of eutrophication, this OSPAR Agreement has set two main objectives: (a) to minimize the human induced eutrophication and (b) to achieve and maintain that all the areas of OSPAR will be non-problem areas by 2020. The OSPAR Commission understands that for these objectives to be effective, a sufficient number of actions is required: to implement monitoring and assessment schemes, to evaluate eutrophic trends in the area (sources and effects), to take into account additional impacts such as hazardous compounds and climate change in marine eutrophication, to identify and propose measures that will fulfill the objectives for good ecosystem health. The Directives incorporating eutrophication components, mentioned above, will be implemented in connection with the OSPAR's objectives. In addition, the effort to reduce nutrient inputs to eutrophic areas by 50 per cent of the levels in 1985 will continue. Any additional measures for combatting eutrophication can be proposed, taking into account their feasibility, cost effectiveness, region specific problems and seasonal factors.

The in depth collaboration among Member States, signatory of the OSPAR agreement, is a good paradigm of improving policy effectiveness. The OSPAR agreement is encouraging the development of interrelations among international institutions for activities linked with marine eutrophication and adopting their measures: measures stipulated by the Directives just mentioned, measures stipulated by the Protocol regarding Emissions of Nitrogen Oxides, measures in the Conventions on Long-Range Transboundary Air Pollution (LRTAP) and measures stipulated by the 73/78 Convention.

Regarding the effectiveness of measures for mitigating eutrophication, positive trends have been observed and significant reduction in nutrient inputs has been recorded (OSPAR, 2017). Between 2008 and 2017, the sea conditions in the eutrophic areas in Denmark, Germany, France, Ireland, Norway and United Kingdom have improved. The overall spatial decrease of eutrophication according to the 2017 assessment is impressive: 169,000 km² of eutrophic areas in 2003, 111,000 km² in 2008 and 100,000 km² in 2017. Eutrophication status mainly improved in the outer and offshore coastal waters of the Greater North Sea, whereas limited improvement was observed in the inner coastal and inshore waters. Mitigating eutrophication levels is largely dependent on the reduction of nutrients of anthropogenic origin discharged into the marine environment. The Convention of Long-Range Transport of Air Pollution has resulted in a reduction of up to 50 per cent in many areas since 1990. In the Greater North Sea, nitrogen inputs have been reduced by 25 per cent whereas atmospheric nitrogen inputs have been reduced by 30 per cent since 1990.

Although the Contracting Parties have achieved a more or less overall decrease in nutrient discharges, this success is not reflected in a relevant decrease in the phytoplankton biomass. This is possibly due to the fact that there is a lag phase between nutrient reduction and limited phytoplankton growth, measured as chlorophyll concentrations, species abundance or recovery of biodiversity. In addition, even the reduced nutrient concentrations (nitrogen, phosphorus) may be sufficient for maintaining high phytoplankton biomass in the water column.

4.7 Eutrophication Status of the East Asian Seas

4.7.1 *Physiography of the East Asian Seas*

The East Asian Seas is a complex system of many seas and straits (Figure 4.11). These include the South China Sea, the Philippine Sea, the Timor Sea, the Sulu Sea, the Celebes Sea, the Arafura Sea, the Banda Sea, The Flores Sea, the Andaman Sea, the Strait of Singapore and the Strait of Malacca (UNEP, 2008). The main seas among the water bodies mentioned above are the South China Sea, the Philippine Sea, the Java Sea and the Arafura Sea. East Asia is characterized by intense economic growth which is interconnected with serious environmental problems including marine eutrophication. The combination of high population and economic growth explain the severity of the serious impact on the marine environment, in some cases destructive. The principal characteristics of the main seas mentioned above will be briefly described, outlining features connected with eutrophication.

The South China Sea, a marginal sea of the Pacific Ocean, encompasses an area of about 3,500,000 km². It communicates with the Pacific in the North through the Taiwan Strait, to the East through the Luzon Strait and the Sula Sea, whereas in the south is extended as far as the Java Sea. The communication of the South China Sea

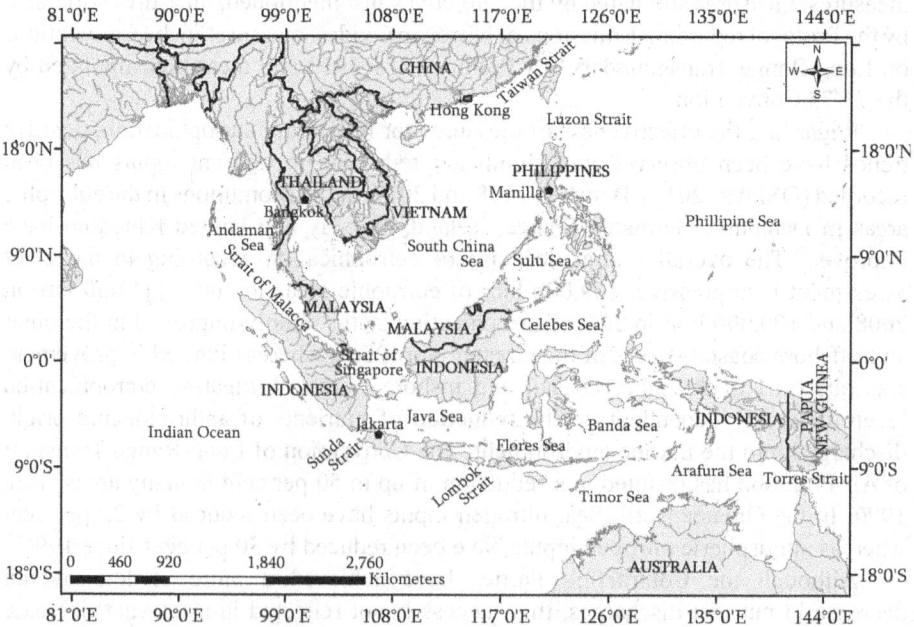

Figure 4.11 The South East Asian Seas.

with the Indian Ocean is through the Strait of Singapore and the Strait of Malacca. The South China Sea also encompasses more than 250 atolls, shoals, reefs, small islands and sandbars. This sea is the second most used sea lane in the world as at least 50 per cent of the international ship tonnage passes through the Strait of Malacca, the Lombok Strait and the Sunda Strait. Among the cargo passing through, 10 million barrels of crude oil per day have to be mentioned. It has been estimated that this water body accommodates one third of the entire marine biodiversity, making the South China Sea ecosystems distinctive. However, fish stocks have been practically depleted as overfishing in the area has already exceeded the maximum sustainable yield (UNEP/COBSEA, 2010).

The Philippine Sea is also a marginal sea occupying a surface area of about 5,000,000 km². The sea is characterized by a degree of complexity and is bordered in the North by the Pacific Ocean, the Philippine Archipelago in the southwest, the Halmahera in the southeast, the Marianas in the North West and Taiwan in the west. Ecosystems in the Philippine Sea are of exotic type. About 500 species of corals occur in the coastal waters. A percentage, about 20 per cent of the worldwide shellfish species appear in the Philippine waters. There are also sharks, moray eels, sea turtles, sea snakes and octopuses in the area. In addition, many species of fish including tuna are commercially caught in the area.

The Java Sea is a rather extensive shallow sea with an average depth of 46 m, covering an area of about 1,790,000 km². It is bordered in the north by the Borneo Island and by the Java Island on the South. Communication of the Sea to northwest is through the Makassar Strait and to the east through the Flores Sea. Between Malaysia and Borneo there is an opening, pretty wide which is the natural border with the South China Sea. There are more than 3,000 of marine species in the area. National parks also exist, the most known being the Karimunjawa.

The Arafura Sea, the most south among the East China Sea, is bordered by the Indonesia Guinea in the north and Australia in the south. It communicates with the Pacific on the west through the Torres Strait. The Arafura marine area is among the richest marine fisheries fields in the world. The main fishery resources are shrimps and venous demersal species.

4.7.2 Environmental pressures and sea conditions in the East Asian Seas

As more than 60 per cent of the inhabitants of the East Asian Region live in coastal cities, there is a serious impact from: (a) land-based activities: these include domestic sewage runoff, industrial waste runoff, poor agricultural practices, tourist development, urban development and uncontrolled aquaculture practices, just to mention the most significant pressures. In addition, land reclamation and watershed modifications cause permanent environmental problems (UNEP, 2008). The population in the coastal cities increases rapidly due to high birthrates as well as due to immigration from inland rural areas. Bangkok, Jakarta and Manilla are a good paradigm of big cities exercising pressures in the marine environment. In addition to the pressures mentioned above, infrastructures like super container ports, airports, new industrial units, expressways, pollute beyond maximum allowable health standards. The marine environment pays a high toll for the economic growth of the area: pollution levels increase by 10–20 per cent every year (Douglass, 2000). Trade and shipping pollution are also sizable sources of pollution although it is more difficult to quantify (ESCAP, 2009). It has been reported (DiMento and Hickman, 2012) that according to a director within the UNEP, the prevailing attitude in the area is the "cowboy mentality" as these countries are prioritizing development and growth at the expense of environmental quality and ecosystems' health.

Poor agricultural practices pollute by discharging their waste in the marine environment that includes herbicides, biocides, pesticides, fertilizers and antibiotics. Tourism in the area increases by millions of visitors every year: facilities characterized by poor design, also enhance pollution problems. However, the major pollution problem comes from sewage disposal into the sea. The whole socio-economic situation for the majority of people that is overcrowding, inadequate sewage systems, slum settlements and human feces contributes not only to marine eutrophication but also cause health problems (Marcotullio, 2001). Land sources contribute 77 per cent to the marine pollution (UNEP, 2000). Rivers carry waste into the sea from inland; this waste accounts for about 70 per cent of the contaminants. Poor management of this type of waste increases marine eutrophication, red tide occurrences and decrease of the fish stock. In addition to this type of point pollution, there is non-point pollution from the runoff of the terrestrial environment and transboundary pollution due to water current and marine hydrodynamics in general. Atmospheric deposition of nutrients has also to be taken into account. However, the last three sources of nutrients are difficult to be quantified.

4.7.3 *Eutrophic conditions in the East Asian Seas*

Eutrophic hot spots in the South East Asian Seas, are usually located near coastal cities, lacking proper sewage systems as well as in estuarine areas. This is a very common problem in China, Indonesia, Thailand and the Philippines (Figure 4.12). The contribution of nutrients into the coastal Chinese waters has not been assessed so far but it is known (Seitzinger et al., 2005) that the global average is about 65 per cent of DIN and DIP. This could be an acceptable approximation also for the rivers discharging into the South East China Sea. Beyond the excessive growth of phytoplankton, outbursts of algal blooms are very common in those areas, fairly often dominated by toxic phytoplankton (UNEP/COBSEA, 2009). Another cause of eutrophication is connected with the rapid development of aquaculture. In addition to the effects of algal growth, fish cultures also contribute to eutrophic conditions.

Seagrasses are widely spread worldwide but the exact area these ecosystems occupy is not known as yet. From rough estimates, seagrass formations in the South East Asia, account for 72 per cent of the global seagrass area. The major economic significance of the seagrass formations stems from the fact that they provide good nursery grounds for many kinds of fishes of commercial interest. They form a source of food for green dugongs and juvenile turtles (Chou, 1994). Eutrophication is considered as the greatest threat to the seagrass ecosystem, South East Asia being the most threatened.

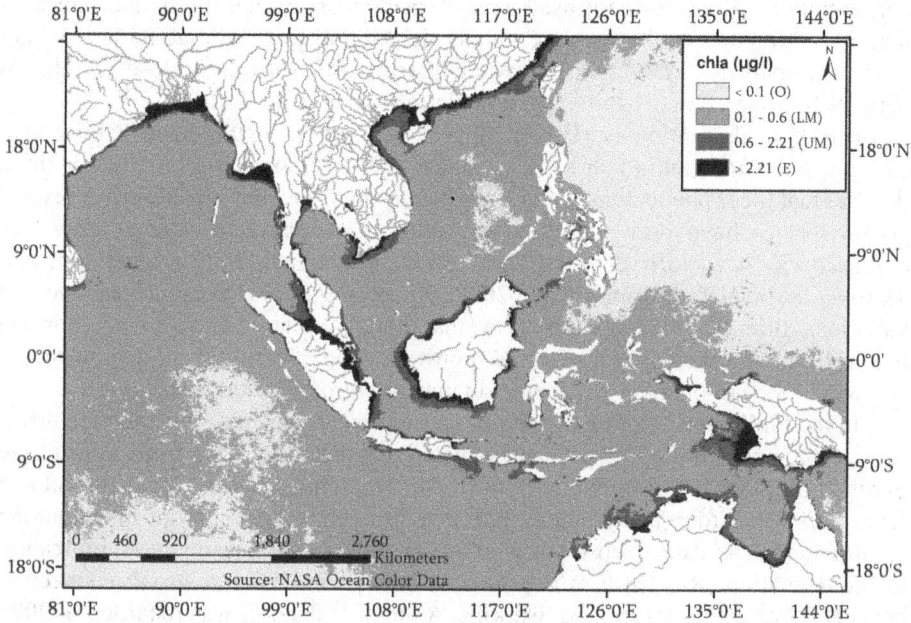

Figure 4.12 Satellite surface chlorophyll distribution in the South East Asian Seas during 2017. Eutrophication scaling according to Simboura et al. (2005-modified) scale.

Chinese rivers account for about 5–10 per cent of the global freshwater input into the sea and about 15–20 per cent of sedimentary material. River water is usually rich in inorganic nitrogen (DIN) and poor in dissolved inorganic phosphate indicated by the high DIN: PO_4^{3-} ratios that varied between 24 and 2,000, spatially and seasonally. River nutrients in the Chinese rivers are enriched from agriculture, aquaculture, urbanization, erosion and climatic conditions (Liu et al., 2009). There is a seasonal trend on river nutrients, as nutrient concentrations in the summer are 3–4 times higher compared to nutrient concentrations in winter. Urbanization, that is sewage discharges, has changed the nutrient regime and silicon has become the limiting nutrient instead of nitrogen and phosphorus. Among the rivers emptying into the South East Asian Seas, 41 per cent has been characterized as polluted whereas 59 per cent as extremely polluted, mainly eutrophic (Smith et al., 2003).

There is no published work on the offshore areas of the South China Sea. Most of the eutrophication studies in the area are confined to estuaries and the coastal waters surrounding the cities. Eutrophication trends over the last 75 years have been assessed in the offshore marine area of Hong Kong, using sediment cores (Hu et al., 2008). Sediment cores can be a useful methodological tool in monitoring historical records resulting from anthropogenic sewage discharge. The results showed that there was an upward trend in eutrophic conditions from 1925 to 2000 in the marine area surrounding Hong Kong. Although the development of eutrophication between 1949 and 1965 was progressive, this rate was accelerated after 1965 and up to the year 2000. Data on eutrophication covering a period from 1985 to 2000, also provide an explanation for the frequent occurrence of algal blooms and red tides. The concentrations of coprostanol used as a biomarker of sewage discharge, indicated that the eutrophic conditions were mainly due to anthropogenic activities. A long term data analysis (1997–2006) in a neighboring area showed a close correlation between sewage effluents and oxygen depletion (Yuan et al., 2016).

Harmful Algal Blooms (HABs) appear very often in the South China Sea, causing, *inter alia*, mass fish kills in aquaculture. It is over the last two or three decades that these phenomena occur at an increasing frequency and severity. Several HABs species have been identified including *Noctiluca scintillans, Pyrodinium bahamense, Phaeocystis globosa, Scrippsiella trochoidea, Heterosigma* spp. and *Mesodinium rubrum* (Wang et al., 2008; Xu et al., 2017). Seasonal and annual variations differ, as it is known by now, that there are many factors involved in the bloom formation including a reversed monsoon wind, river discharges, upwellings and winds appear to be the key factors in marine eutrophication in the area.

In the Philippine Sea, the two major causes for marine eutrophication are urban sewage and aquaculture. It has been estimated that fish production from aquaculture in the Philippines Sea is about 2.48 million metric tons per year. Uneaten food and feces from fishes form a substantial source of nitrogen and phosphorus in the marine environment. The daily feed input which is about 2 kgm^{-2} contributes significantly to sedimentation (Reichardt et al., 2006). There seems to be a strong connection between toxic algal blooms and fish kills. A massive fish kill was recorded in 2002 alarming authorities, stakeholders and scientists for improving water quality as far as the eutrophic conditions are concerned. This fish kill coincided with the bloom of the dinoflagellate *Prorocentrum minimum* (Lourdes et al., 2008). In spite of the measures taken, the coastal waters in the area remained eutrophic for ten years, after imposing regulations possibly due to uneaten and undigested feed deposited in the sediments (Ferreira et al., 2016). Harmful algal blooms have also been reported in Malaysian waters (Tee, 2012).

4.7.4 Policy effectiveness in the East Asian Seas

In spite of the vast area covered by the South East Asian Seas, the high population densities in many cities, the growing economic activities in the area, the causes/effects in the quality of the marine environment and ecosystems' health, there is relatively limited information and publications in international journals on the conditions of the sea and marine environmental quality. Most of the publications regarding eutrophication refer to specific local problems such as sewage eutrophication from large cities, cultural eutrophication from land farming and mariculture as well as impact of eutrophic conditions on the coral reefs, coastal waters and the decrease of the fish stocks. Observations at a large scale, covering offshore areas are based on the East Asian Seas Environment Outlook (EASEO) but the data are still in the process of analysis. Efforts have been made to collect information within the Global Ocean Observation System (GOOS). This system is using information from satellites, buoys and other devices that can collect relevant information (DiMento and Hickman, 2012). Under those conditions, it is difficult to come to a reliable assessment of the field conditions which would allow the authorities involved to evaluate the effectiveness of the measures taken by the states that participate in "*The Action Plan for the Protection and Development of the Marine Environment and Coastal Areas of the East Asian Region (the East Asian Region Action Plan)"* that was initially signed in 1981 by Indonesia, Malaysia, the Philippines, Singapore and Thailand. COBSEA is the "Coordinated body on the Seas of East Asia" established in 1993; it functions as the Secretariat of the COBSEA. The concept of activities regarding the South China Seas is focused by UNEP and GEF on *"regional cooperation in the identification of sensitive ecosystems, land based contamination problems, evaluation of their significance and development of standards for national level adoption within a regional context in order to develop an approximately precautionary approach to discharges to the East Asian Sea marine basins"* (UNEP, SCS, 2008). Among the objectives set by the COBSEA, the assessment of the land-based activities on environmental quality is closely connected with eutrophication problems; within the COBSEA objectives is also the development of measures towards the effectiveness of the Action Plan. The focus of COBSEA on land based pollution will mainly focus on nutrients, sediments and wastewater; the purpose of these initiatives will be to *"prevent and reduce eutrophication and sedimentation and their impacts on the marine and coastal environment"*.

The unsatisfactory effectiveness of the measures (UNEP/COBSEA, 2018) is due to sectoral policies, inadequate planning for coastal and marine space and inefficiency in pricing environmental economic data such as ecosystem services' values. Data sharing also remains limited. Among the measures to be taken will be source identification, combined with prevention, reduction and control of nutrient discharges. In addition, emphasis will be placed on information exchange, best practice implementation as well as technical training and infrastructure development.

From the point of view of effectiveness in improving environmental quality, the results are not encouraging. May be that economic growth in most of the South East Asian States is a priority compared to environmental issues. The science component of the East Asian RSP does not seem to provide the quality and quantity of information characterizing other regional seas such as the Mediterranean Sea, the Baltic Sea and the North Sea. However, there is some progress; a professional from the COBSEA Region (DiMento and Hickman, 2012) commented that the biggest contribution of the East Asian RSP is adding some "science" to the system.

The effectiveness from the point of view of improved relations among States, the institutions forming the South East Asian cluster and people is more encouraging mainly for three reasons: (a) the biggest countries in the area are signatory parties in the Action Plan (b) the COBSEA secretariat organized collaboration among the signatory states and (c) most of the Member States of the COBSEA showed an interest in implementing marine agreements regionally. However, the drawback is the fact that there is no convention in the area, legally binding the member states about their responsibilities within the South East Asian RSP framework.

The lack of a legally binding agreement is the reason that few effective outcomes have resulted, reducing the effectiveness indicated by the implementation of environmental law and policy. This, in spite of the fact that some of the COBSEA Members are legally bound to international conventions connected to marine environmental protection and management. International conventions with eutrophication component signed among COBSEA States are the *"Protection of the*

Marine Environment from Land-Based activities". The Convention for the "*Control, Management of Ships Ballast Water and Sediments*" and the MARPOL Convention for the prevention of pollution from ships.

In conclusion, lack of a binding agreement, deficiencies in capacity and regional cooperation, limited funding and jurisdictional confusion are the main reasons for the limited success regarding marine environmental issues. Deterioration of the South East Asian Seas, cannot rely any longer on national willingness. The problem of eutrophication seems to be the principal problem in the area and the states are generally reluctant to be effective, in spite of the measures agreed, due to the cost of the measures but also because of the restrictions that have to be set defining the limits of some activities.

4.8 Eutrophication Status of the Wider Caribbean Region

4.8.1 *Physiography of the Wider Caribbean Region*

The Wider Caribbean Region (WCR) is a marine system that comprises of the Caribbean Sea, the Gulf of Mexico and the neighboring bays and marine areas (DiMento and Hickman, 2012). The WCR covers an area of about 15,000,000 km² (IOC-UNESCO, 2009). Due to this great size, the WCR is divided into four subsystems (NOOA, 2011): the Caribbean Sea LME, the Gulf of Mexico, the Southeast US Continental Shelf LME and the North Brazil Current LME (Figure 4.13). This division system also has been adopted by the Global International Waters Assessment (GIWA) and the Large Marine Ecosystem (LME) classification scheme.

The Wider Caribbean Sea (CS) with an area of 1,063,000 km² and is part of the Western Atlantic Ocean. The Caribbean Sea is surrounded to the south by Venezuela,

Figure 4.13 The Wider Caribbean area.

Colombia and Panama, to the west by Costa Rica, Nicaragua, Honduras and Guatemala as well as the Yucatan Peninsula of Mexico. Big rivers including Amazon, Oniroco and Magdalena discharge into the Caribbean. Due to the rivers and currents, pollutants tend to be dispersed in the area. There is also an inflow of subsurface water into the Caribbean across two sills: the Anegada Passage (approximate sill depth 2000 m) is located between the Virgin Islands and the Lesser Antilles, whereas the Windward Passage (sill depth about 1,600 m) is between Cuba and Hispaniola. North Atlantic deep water masses enter the Caribbean through the Windward Passage; their salinity is 35 psu and their oxygen content is high. The climate of the Caribbean is generally tropical but there are local variations depending on mountain elevation, trade winds and water currents.

Coral reefs in the Caribbean cover 50,000 km² and account for 9 per cent of the world's coral reef area. The coral reefs encompass 70 species of hard corals and are the habitat to more than 500 species of fishes. In addition to the coral reef formations, many lagoons provide favorable conditions for seagrasses. Among the seagrasses, dominant species are *Thalassia testudinum*, and *Syringodium filiforme*. There are also black and red mangrove formations in the form of dense forests surrounding lagoons and estuaries. The presence of about 1,000 species of fish has been documented in the Caribbean, including sharks, flying fish, manta rays and angel fish. There are also 90 species of sea mammals in the area: the most common are whales, dolphins, seals and manatees.

4.8.2 *Environmental pressures and sea conditions in the Wider Caribbean Region*

The population of the states bordering the Caribbean Sea grows continuously and it is expected to be about 90 million by the year 2020 (UNEP, 2006). Most of the population inhabits along coastal areas; the population densities expand over the tourist season. This population increase is followed by rapid coastal development which is the reason of marine environmental deterioration. The WCR is among the largest oil producers worldwide; it is estimated that oil production exceeds 170,000,000 tons of crude oil per year. Oil transportation, offshore production and land-based activities (oil terminals and refineries) contribute to petroleum pollution. It has been estimated that about 80 per cent of the total oil discharged into the marine environment comes from oil refineries. However, oil pollution problems are restricted to coastal waters whereas oil pollution in offshore areas is not alarming.

Solid waste is also considered as an emerging problem; the situation is not easily manageable since land space for dumping in the insular Caribbean is limited. Incineration and recycling (plastics, metals, glass and papers) have already been proposed to mitigate these problems (Siung-Chang, 1997). A UNEP overview on marine pollution has identified 14 sources of pollution from industrial activities since a long time ago (UNEP (OCA)/CAR, 1994). These activities, producing liquid discharges, are food industries, distilleries, sugar refining factories and other chemical industries. But the most widespread pollution problem in the WCR is sewage. This is the main source for marine eutrophication in the area. Eutrophication is also the main threat for the Caribbean coral reefs. Extensive deaths of zooxanthellae, the symbiotic partners of corals that give them the color, have been recorded, causing coral bleaching. About 42 per cent of the corals are bleached and about 95 per cent are experiencing some degree of deterioration. Eutrophication has also been connected to lower recruitment and survival rates of coral species in the western coast of Barbados (Hunte and Wittenberge, 1992).

4.8.3 *Eutrophic conditions in the Wider Caribbean Region*

Eutrophic conditions in the area are due to untreated or inadequately treated sewage (Figure 4.14); these discharges contribute towards excessive nutrients, suspended solids and pathogenic bacteria in the marine environment. Only a small percentage of sewage is treated properly.

Sewage production from livestock (cows and pigs) is also substantial. Algal blooms occur fairly often and seven fish kills were recorded between 1976 and 1990 (Heileman and Siung-Chang, 1990). Sedimentation of organic particles can cause suffocation of hard coral species (Wynne, 2017). Overgrowth of macroalgae can have similar effects (Fabricus, 2005). Cyanobacterial growth is also enhanced under eutrophic conditions; this has been connected to coral diseases, since it is supported that cyanobacterial growth favors coral diseases (Wynne, 2017). Cyanobacteria have also negative effects on sponges (Butler et al., 2005).

A recent study on eutrophication in the Caribbean (Wynne, 2017) has brought evidence that in addition to land-based sources of eutrophication, including significant nutrient sources from densely populated urban areas of South America and the Gulf of Mexico States, nutrient sources originated from the Amazon and Orinoco, can be as powerful in supporting eutrophic trends as the local sources. Deforestation in the Amazon River results in runoff that has increased significantly the levels of nutrients

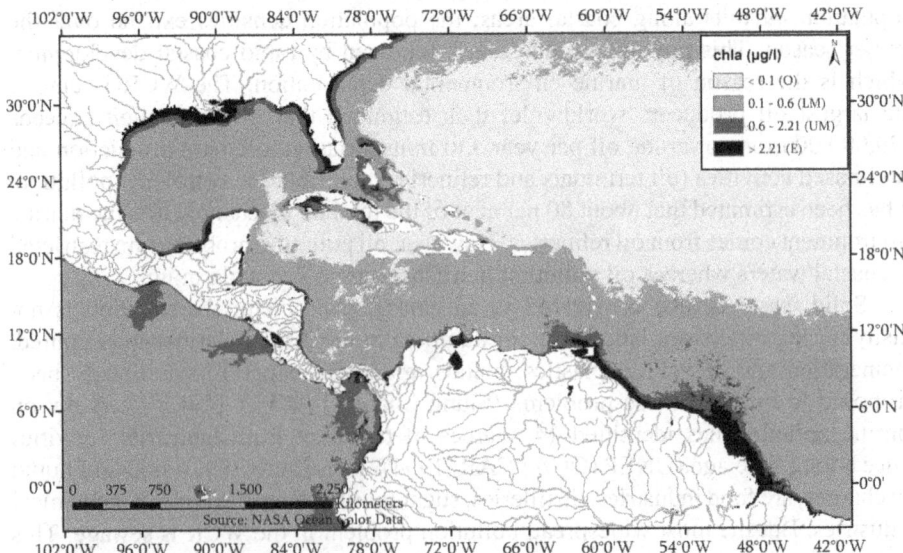

Figure 4.14 Surface chlorophyll distribution in the Wider Caribbean Region during 2017. Eutrophication scaling according to Simboura et al. (2005-modified) scale.

in the plume (Santo et al., 2008). It has also been supported that water current patterns either recirculate nutrient-rich waters or inhibit departure of nutrient-rich water masses from the area. The secondary and tertiary effects of these nutrients, can act synergistically with other stressors, mentioned in the previous section, causing habitat degradation and eventually bad ecosystem health and biodiversity losses. Recent studies have indicated that anticyclonic movements, when occur, introduce more nutrients in the Caribbean, originally from the Venezuelan coast. Gyres, the result of recirculation, inhibit the loss of productive waters through the Yucatan Passage. This is a potential mechanism explaining why nutrients are retained and built-up in the area overtime. This nutrient build-up also explains the frequent algal blooms events in the area.

4.8.4 Policy effectiveness in the Wider Caribbean Region

The Caribbean Regional Seas Programme (RSP), known as the Caribbean Environment Programme, is the basic agreement for the marine protection in the Wider Caribbean Region, encompassing 36 states or territories (Colmenares and Escobar, 2002). The Caribbean Action Plan, derives from the Convention for the Protection of WCR. The activities include among a wide number of objectives, control of the marine pollution, evaluation of impacts on the coastal waters, evaluation of natural hazards and protection from marine pollution by Land-Based Sources (LBS). The LBS Protocol consists of four Annexes indentifying activities and pollutants, domestic wastewater issues and agricultural non-point sources of pollution. Although not explicitly reported, within the targets of these Annexes, nutrient and eutrophication are included *de facto*.

The Convention of the Caribbean has been characterized as the most comprehensive RSP; this Convention with its protocols, provides coverage for many problems referring to the marine environment (Sheehy, 2004). The Cartagena Convention is also characterized by a strong cluster: it includes a number of players such as United Nations Food and Agriculture Organization (UNFAO), UNESCO, the Caribbean Sub Commission or IOCARBE (Fanning et al., 2007). There are other Conventions overlapping or supplementing the Cartagena Convention. These include the Convention for the Biological Diversity (CBD), the Framework Convention on Climate Change (UNFCC), the Convention of Wetlands (Ramsar Convention). NGOs also play a significant role in the region.

It is difficult to assess the physical conditions in the area and therefore the effectiveness indicated by physical parameters is limited as well as the production of publications in international journals. There are also shortcomings regarding the law and policy framework. There are gaps on legal mandates, overlaps by different legal instruments and requirements of overgeneralization. These gaps refer to mandatory processes and measures for prevention or mitigation of marine eutrophication.

In the effectiveness indicated by improving relations among the Member States, it looks as if there is a degree of success. The main objective of CEP is to build capacity, disseminate scientific information and draw management guidelines (CEP, 2010). Also to connect the institutions involved, including NGOs and public participation. This has been the case from the beginning as public participation played a significant role in the drafting of the Cartagena Convention. They were open meetings where participants, in addition to the public, were NGOs and private business partners. This allowed the interesting parties to negotiate many issues and include them in protocols and annexes.

In conclusion, the CEP has been partly successful. Furthermore, the WCR marine environment is a vital economic asset in the area. Among the economic activities, tourism seems to be driven away, accounting for a great deal of the revenue in the area. As visitors are attracted by the beauty of the natural marine environment, good ecosystems' health is a "must" and therefore prevention of eutrophic conditions, among other measures, is a prerequisite for the welfare of the area.

4.9 Eutrophication Status of the West and Central African Seas

4.9.1 *Physiography of the West and Central African Seas*

The region of West and Central African Seas known as WACAF is made of 22 states, most of them coastal (Figure 4.15). Guinea, Sierra Leone, Liberia, Cote d'Ivoire, Ghana, Togo, Benin, Nigeria, Cameroon and Gabon are the most significant regarding population, economic activities and impact on the marine environment (UNEP/GPA, 2005). There are four narrow basins of sedimentary origin with a strong influence on the pattern of river drainage. The main river systems in the area are four, spaced out

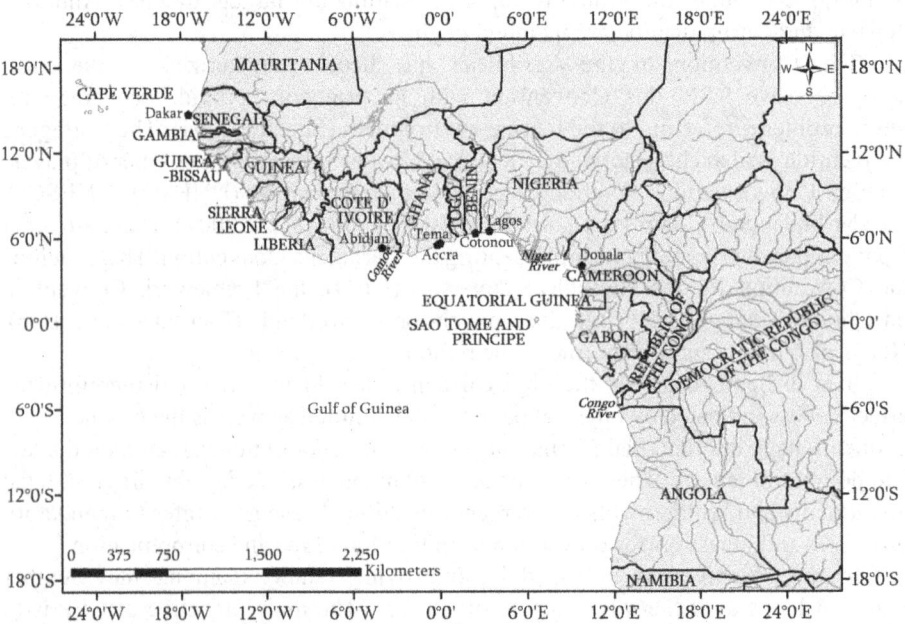

Figure 4.15 The West and Central Africa Marine Area.

from Senegal to Congo (UNEP, 1999). The most important river is the Niger River that drains an area of about 1,000,000 km². The Volta River is draining an area of about 390,000 km². The Congo River is the second river regarding the catchment area in the world; the Comoe River in the Cote d'Ivoire is also considered as a large river in the area. Dams have been along these rivers for irrigation, energy production and flood control purposes. Various fresh water management policies in the region imply downstream impacts including the acceleration of coastal erosion processes and coastal hydrology. On the contrary, in the offshore areas of the Gulf of Guinea, there are five distinct currents that are relatively persistent. These are important from the point of view of transportation of substances, temperature stability and the biological regime. Briefly, these five currents are: (a) the Benguela current flowing along the coastal areas of Namibia and Angola (South West African coastal zone) (b) the Guinea current that carries warm waters along the coast of Guinea near the Equator, as it moves eastwards and southwards (c) the Equatorial Countercurrent which is the continuation of the Guinea current (d) the South Equatorial current that flows between 10°S and the Equator, some distance from the coast and (e) the Canary Current in the Northern Part that feeds the Guinea Currents as well as the North Equatorial Current. The Canary Current also drives an almost permanent upwelling mainly along the coasts of Gambia, Mauritania and Namibia. These cool, nutrient rich upwellings have profound effects on the productivity of the nearshore marine environment. The combination of high precipitation and significant riverine, freshwater outflow keeps the relatively warm coastal waters above 24°C and salinity values below 35 psu.

The coastal morphology from north to south is dominated by (a) arid and sandy coastal plains bordered by dunes (b) sandy, marshy estuaries and deltas with mangrove vegetation and (c) extensive marshy areas formed by the Niger Delta. In addition, there are islands and an archipelago off the coast of West Africa (Canary and Cape Verde Islands) and in the eastern part of the Gulf of Guinea (UNEP, 2006).

A wide variety of natural habitats and ecosystems exist in the western and central coasts of Africa. The most important among them are: (a) Wetland habitats: these are extended from Senegal to Angola over a stretch exceeding 25,000 km. These mangroves, although not particularly diverse, compared to the mangroves of the Eastern part of the continent, are the most well developed in Africa. They function as a nursery for many fish and shellfish species and provide exquisite ecological conditions and habitats for many species and birds (b) coastal lagoons mainly from Cote d'Ivoire to Nigeria, connected with estuarine systems, deltas and freshwater water masses (c) some seagrass beds and (d) sandy beaches: they are mainly extending along the Mauritania and the northern Senegal coasts. They are ecologically important as nesting sites, especially for the sea turtles.

The diversity in the coastal waters is rather high, with 239 species of fishes having been recorded, many of these with commercial interest. There are also large populations of resident and migrant seabirds and marine mammals such as manattee's (*Tricherus senegalensis*) occurring in some lagoons.

4.9.2 Environmental pressures and sea conditions of the West Central African Seas

The population of coastal states of the WACAF region that inhabits a coastal marigin 60 km wide is difficult to be estimated but it may exceed 60 million inhabitants. There are key cities where population density increases continuously due to a high growth rate and constant movement of people from inland rural areas. The most populated coastal cities are Accra-Tema, Abidjan, Cotonou, Dakar, Douala and Lagos. Lagos is a good example of exponential population growth: from a city of 350,000 inhabitats 30 years ago, has reached 7,000,000 people and is still growing (Kouassi and Biney, 1999). It has to be taken into account that Lagos accommodates half of the manufacturing capital and almost half of the skilled manpower in Nigeria. The cities of Abidjan and Accra-Tema accommodate half of all the industries of the Ivory Coast and Ghana, respectively.

This fast growing population, where the wide majority of people live under conditions of extreme poverty causes a number of environmental pressures and finally problems, mainly due to raw sewage that enters the marine environment directly through sewage system pipelines. The most important effects from sewage eutrophication are public health problems due to unsuitable drinking water and contaminated bathing waters. Industrial growth acts in a synergistic way with impacts from the city life: disposal of hazardous chemicals of untreated industrial effluents released into the marine environment and atmospheric deposition of gaseous and particulate noxious compounds are the primary causes of industrial pollution. That means loss of biodiversity, water quality decline and impacts on biological resources.

The worst deficiencies in environmental quality are connected with diarrhea and malaria in densely populated areas where low income people live. According to Koussi and Bineay (1999): "*Poverty itself is a contributing factor to the present state of degradation, since it is a strong impediment to adopting new practices or behavior less damaging to the environment*". Solid waste, mainly litter and plastics, are also major contaminants. This problem is a nuisance in swimming beaches and coastal residential areas. In addition, there are metal cans scattered, abandoned pieces of fishing gear and fishing vessels as well as logs carried into coastal areas through river water (Kouassi et al., 1995). It has been estimated that 80 per cent of marine pollution in Africa comes from land-based sources with a significant contribution of untreated sewage and fertilizers (UNEP, 2013). In Nigeria, 20 million people that is 23 per cent of the total population lives along the coastal zone; this proportion can vary between 30 and 90 per cent of the population of the states. According to the existing data (Scheren and Ibe, 2002), population growth in the areas is about 3 per cent and the probable doubling time is about 25 years. This growing population applies environmental pressures in a twofold way: (a) over-exploitation of the natural resources in a non-sustainable way and (b) marine pollution especially in the inshore areas. The most important environmental pressures are shortly presented in the following paragraphs.

The major cities along the coastal zones are principal commercial and industrial centers. Abidjan is a good example: the city hosts 70 per cent of all industrial activities in Cote d'Ivoire; this accounts for 80 per cent of the total industrial production of the country. Industries in the coastal zone of Cameroon accounts for 60 per cent of the national production. These rates for Togo, Benin and Ghana are 85, 80 and 65 per cent, respectively. Industrial activities are either oriented towards the needs of local communities (i.e., food, beverage, textile, soap and detergent production) or processing of raw materials (petroleum in Nigeria, phosphate in Togo, aluminum in Ghana). Problems from waste discharges are beyond control, leading to environmental degradation and deprivation of the local population from vital resources (drinking water and food).

Exploitation of mineral resources causes a number of environmental effects. Extraction of sand and gravel causes coastal erosion. Oil exploitation, especially in Nigeria is connected with consequent oil spills and blowouts, devastating coastal areas by destroying marine habitats and causing mass fish mortality. In addition, mining of metals such as gold in Cote d'Ivoire is leading to soil degradation and furthermore runoff as well as leaching of nutrients from these sites affects lagoon water quality and coastal eutrophication. Among the chemical pollutants, waste oil of industrial origin is also a sizable source of degradation of the marine environment. There are also residues such as phosphate, mercury and zinc detected in the sea water (Heileman, 2009).

The dominant economic activity in the WACAF States, apart from Nigeria, is agriculture. Cultivation of food crops (maize, rice and vegetables) and cash crops (oil palm, cotton, rubber and pineapples) is widespread around rivers, wetlands and coastal zones. Uncontrolled use of fertilizers and pesticides adds toxic pollutants to the marine environment. There are also pollutants in the marine environment stemming from forestry. Most of the trees providing commercially valuable wood have been cut; this has caused severe degradation of soils easing runoff and leakages of nutrients into the marine environment, contributing to eutrophic trends. As many households are still depending on firewood for cooking, many mangrovial forests have been sacrificed for domestic use of the wood, affecting marine biodiversity and exposing the shores to erosion.

As tourism, especially along the coastal zones of the WACAF Region is growing exponentially, in some of the states like in Ghana, the impact on the marine environment is multi-facet: litter and marine debris have an obvious impact on the landscape aesthetics of the coastal zone, whereas sewage pollution affects coastal water quality.

4.9.3 Eutrophic conditions in the West and Central African Seas

Eutrophication is a coastal problem in WACAF, originating from land-based sources that have already been mentioned above (Figure 4.16). These are urban domestic facilities, non-point runoff, industrial facilities and mining sites. Urban sewage sources include waste treatment plants, pipelines discharging raw sewage, markets, households, schools and hospitals. The main industrial sources contributing to eutrophic conditions are discharges from breweries, food factories, textile industries and wood processing plants.

Nutrient leaching and processes are difficult to quantify for a number of reasons: missing information in many areas and poor time series; extremely dynamic spatial and

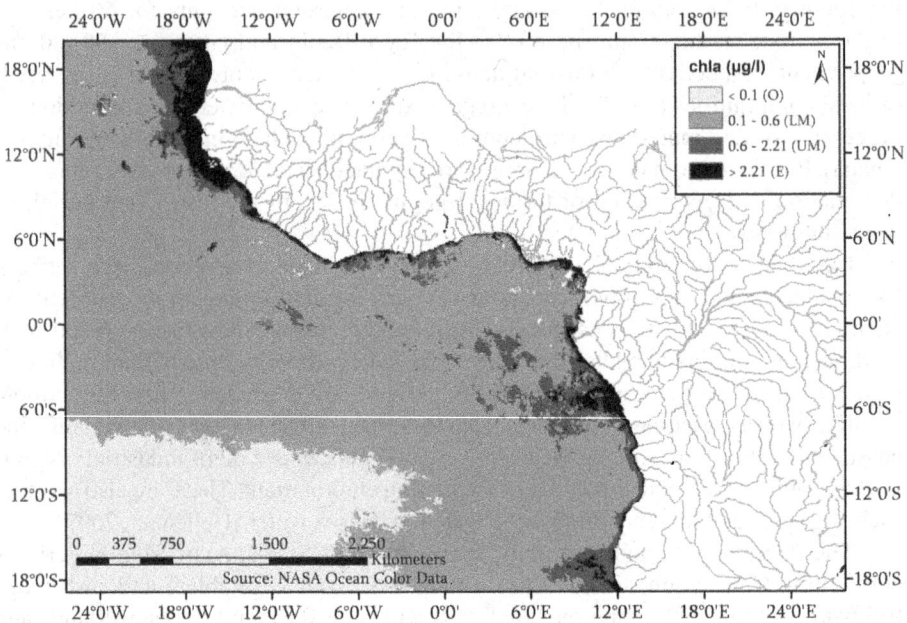

Figure 4.16 Surface chlorophyll distribution in the West and Central African Seas during 2017. Eutrophication scaling according to Simboura et al. (2005-modified) scale.

temporal variability; complicated hydrology and temperature variations in the Gulf of Guinea. However, it has been estimated that the hydrologic export to the Atlantic is about 1.5×10^9 kg.y^{-1} whereas losses from de-nitrification are about 1.1×10^9 kg.y^{-1} (McGlade et al., 2002). Enrichment of the surface layer (0–30 m) with phosphates observed between mid-December to mid January was recorded when a transient intrusion of water masses occurred, rich in phosphates (Arfi et al., 2002).

The information on plankton is even less. A solution has been given to an extent by using "ships of opportunity", that is commercial vessels that deploy Continuous Plankton Recorders (CPR) fitted with sensors to measure temperature, salinity, nitrate, nitrite and chlorophyll as well as other parameters not directly connected to phytoplankton and eutrophication (Sherman and Anderson, 2002). Microscopical examination of samples from CPR in December 1995 from the eastern Gulf of Guinea and south of the Niger Delta have shown that the phytoplankton community was dominated by *Rhizosolenia calcar-avis*, *Rhizosolenia alata*, *Chaetoceros* ssp. and *Navicula* ssp. The dominant taxa along the coastal waters of Ghana between Takoradi and Cape Three Points were *Chaetoceros* ssp., *Bacteriastrum* ssp. and *Trichdesmium* ssp. The cyanophycean *Trichodesmium* spp. was also abundant between the San Pedro area (Cote d'Ivoire) and Takoradi (John et al., 2002).

Eutrophication and oxygen depletion is a phenomenon often occurring in the lagoon of the Gulf of Guinea. Assessment of eutrophication in six countries of the region (Scheren et al., 2002) has shown that organic pollution in the coastal lagoon studied, apart from oxygen depletion, decreased fish landings and enhanced outbreak of waterborne diseases. Eutrophication is also having negative effects on the mangrove systems of the Gulf of Guinea. The economic and ecological importance of the mangroves is well known for the ecosystem services they provide that include carbon storage, nursery support for fisheries, shoreline building, forestry products and recreational opportunities (McLeod et al., 2011; Koetse et al., 2015). In the mangroves of Malanza (Sao Tome Island), relatively high chlorophyll concentrations were recorded; pigment ratios showed dominance of diatoms and green algae. Among the diatoms, pennates seemed to have a prominent presence, *Navicula* spp. being a commonly occurring species. Roughly speaking, phytoplankton growth was found to be at acceptable levels, a fact that was also supported by the application of the TRIX Index (Brito et al., 2017).

Work on harmful phytoplankton species in the area is limited. A preliminary investigation along the Nigerian coasts was carried out between November of 1999 and April 2011; this work provided indications about the presence of 18 potentially harmful dinoflagellates; these included the genera Ceratium, Dinophysis, Gonyaulax, Gymnodinium, Lingulodinium, Prorocentrum and Scrippsiella (Ajuzie and Houvenaghel, 2009). The presence of the nitrogen fixing cyanobacterium Trichdesmium spp., has been reported in the northwestern African regions (Hood and Coles, 2004; Tyrrell et al., 2003).

In conclusion, published work on phytoplankton ecology and eutrophication in international journals about the Gulf of Guinea is scanty. The emphasis of the existing publications on marine research in the area has been placed on physical oceanography and fisheries.

4.9.4 Policy effectiveness in the West and Central African Seas

The main legal tool in the area is the Abidjan Convention and its two protocols for environmental protection, including pollution, coastal erosion, mammals and environmental impact assessments. However, the large number of the Signatory States, their national legislations, and the numerous institutions with jurisdiction that overlap in management and coastal issues, create conflicts about policies and plan implementation (Kouassi and Biney, 1999). Due to inadequate financing, lack of skilled manpower in many cases and inadequate equipment, there is a limited amount of data, environmental impact studies and publications. This makes both management and policy making difficult, introducing a degree of uncertainty due to weak documentation. An additional problem, is the unwillingness of most of the States to accept the cost of the measures to be taken for marine environmental protection, in spite of the long term profits. Once they have to decide between marine environmental quality and growth, growth is usually the winner.

Under those conditions, that is shortage of funding, shortage of experienced personnel and absence of adequate information, it is difficult to rank environmental problems. However, based on the existing information, sewage of domestic origin, its effects on public health and marine ecosystem's quality, seems to be at the top of the list. The conditions regarding coastal eutrophication need to be improved. This will be achieved if the attitudes change and more scientific assistance will be provided.

The effectiveness improved relations among states, institutions and people seem to have some positive effects. The Abidjan Convention encourages the collaborations among states and institutions. The cluster of actors in the WACAF area has increased progressively since the enforcement of the Convention and the onset of the UNEP WACAF Regional Sea Programme. The cluster includes many international and multilateral organizations such as the International Maritime Organization (IMO), the Food and Agriculture Organization (FAO), the International Oceanographic Commission (IOC), the World Health Organization (WHO) and the International Atomic Energy Agency (IAEA).

In conclusion, the WACAF Programme instigated the awareness among people and decision makers and led government agencies to accept responsibilities for the protection and management of the coastal areas. The issue of eutrophication should be added as a principal objective in the protocol(s) and regular assessments of the trophic status of the coastal system and the effects on ecosystems should be evaluated regularly.

4.10 Eutrophication Status of the Gulf (ROPME Area)

4.10.1 Physiography of the Gulf

The Regional Organization for the Protection of the Marine Environment that functions within the Kuwait Convention, covers an area that includes the Persian or Arabian Gulf (hereafter referred to as the Gulf), the Strait of Hormuz, the Gulf of Oman and the South Eastern coasts of Oman located in the Arabian Sea (Van Lavieren and Klaus, 2013). These water bodies are geographically discrete and each of them exhibits distinct physical, chemical and biological characteristics. This ROMPE Sea Area is surrounded by eight states, all Member States of ROPME: the Islamic Republic of Iran, the Republic of Iraq, the State of Kuwait, the Kingdom of Saudi Arabia, the Kingdom of Bahrain, the State of Qatar, the United Arab Emirates and the Sultanate of Oman. The exact geographic coverage of ROPME is provided in the Article II of the Kuwait Regional Convention as follows: 16°39'N, 53°3'3"E; 16°00'N, 53°25'E; 17°00'N, 56°30'E; 20°30'N, 60°00'E; 25°04'N, 61°25'E.

The Gulf is a water body surrounded by Iran to the NE and the Arabian Peninsula to the SW. It communicates with the Indian Ocean through the Straits of Hormuz and the Gulf of Oman. The area of the Gulf is about 251,000 km². Its length is 989 km and the width, in the narrowest point, the Strait of Hormuz, is about 56 km. It is a shallow gulf, the average depth being 50 m and the maximum depth 90 m. On the western end is the Shatt al-Arab estuary which is the common waterway of Tigris and Euphrates Rivers (Figure 4.17).

The Gulf is an evaporation basin with river inflows (about 2,000 m^3s^{-1}) and water from precipitation (about 1,800 mmy^{-1}). In spite of these inflow rates, evaporation is high and there is an estimated water deficit of about 416 km^3y^{-1}. This deficit in water mass is replenished by surface water masses inflowing through the Straits of Hormuz. The outflowing water characterized by higher salinity, follows the way out through the bottom of the Gulf. According to the available data and the published work on the three marine areas of the RSA (ROPME Sea Area), the renewal (residence time) that takes the water mass of the Gulf to be replaced by the inflow waters through the

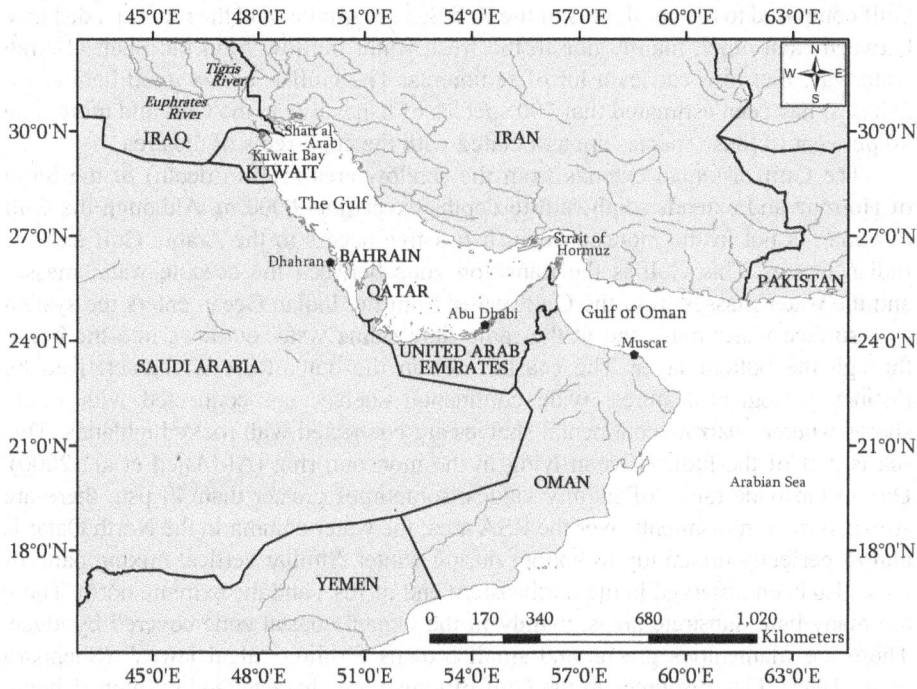

Figure 4.17 ROPME Sea Area.

Gulf of Oman is between 2.1 and 5 years (ROPME, 1988; ROPME, 2013; Reynolds, 1993). These residence time assessments are valuable in estimating the dilution of nutrients and various other pollutants.

   Although life in the Gulf is not as diverse as in the Red Sea, there is high diversity including indigenous species due to hydrological characteristics and to the relative isolation from the Indian Ocean. There are mangrovian ecosystems characterized by a diverse group of trees and shrubs that are dominated by *Avicennia* or *Rhizophora* which form important habitats. Mangrove formations seem to settle in mudflats. They provide biota for more than 2,000 species of fish, invertebrates and epiphytic plants. Their distribution declines and it is estimated that only 130 km² of mangrove formations remain; 80 per cent of those are in the Iranian Side. Genus is Avvicenia marine that can tolerate a wide spectrum of environmental conditions.

   Tidal mud flats are habitats that favor primary productivity. They form the major part of the Kuwait coastal areas and the proximity of the Shatt Al-Arab Delta. Mud flats are dominated by the presence of cyanophyta; beyond their productivity they provide feeding ground for many organisms. The Gulf supports more than 500 species of fish; about 125 species are using reefs as habitats whereas the rest are either pelagic or benthic.

   Among the RSA ecosystems, coral reefs seem to have a prime place. Coral growth in the Gulf is insignificant below the depth of 15 m. In the wider area of the plume of Shatt-Al-Arab, prevailing conditions preclude corals. The coral reefs of the

Gulf compared to the coral reefs of the Red Sea are smaller and the reefs at a distance between each other, mainly due to the fresh water outflow from the Shatt al-Arab waterway, that also carries a lot of sediments. The Gulf is also a good habitat for fishes. It has been estimated that 700 species of fishes live in the Gulf and more than 80 per cent of these species are associated with the coral reefs of the area.

The Gulf of Oman extends from the shallow area (100 m depth) of the Strait of Hormuz and extends southward to depths exceeding 2,000 m. Although the Gulf of Oman is not in the monsoon belt, it has free access to the Arabic Gulf and the Indian Ocean. This Gulf is the transition zone between the oceanic water masses and the water masses from the Gulf: water from the Indian Ocean enters the system as a surface water mass and at the same time saline water outflows into the ocean through the bottom layer. The coastal area in the outer RSA is characterized by distinct geological features: wide continental shelves are connected with sandy shores whereas narrow continental shelves are connected with rocky highlands. This sea is part of the Indian Ocean lying in the monsoon ring (Al-Majed et al., 2000). Due to the wide range of salinity values, sometimes greater than 40 psu, there are strong vertical movements over the RSA area: the water column in the North Qatar is almost perfectly mixed top to bottom during winter. Similar vertical mixing patterns have also been observed in the northeastern end of RSA and the extreme north. There are many hard substrate areas, mainly in the Omani coastal zone covered by algae. These are filamentous greens and small browns forming "algal lawns" (Sheppard et al., 1992). Fish resources in the Gulf of Oman are characterized by high richness and diversity; fisheries is the second natural resource of Oman after oil and natural gas.

Regarding water circulation, there are indications that surface water masses move in a cyclonic way within the RSA area. Wind action and evaporation have been suggested as the main driving forces explaining the circulation pattern. The overall circulation is rather complex and presented in the report by Chao et al. (1992).

4.10.2 *Environmental pressures and sea conditions in the ROPME Area*

Marine environmental quality in the area is dependent on both land-based and sea-based activities that exercise pressures on the water quality and ecosystems' health. Pressures stem from the population growth in the area: the total population increased from 46.5 million in 1997 to approximately 150 million in 2010 (ROPME, 2013). Urban areas have some effects on the marine environment mainly through sewage discharges. Industrial developments in the RSA area have been unprecedented over the last 40 years; the most common heavy industries on the coastal zone are power plants, desalination units, refineries and petrochemical factories. Light industry includes food and beverage factories as well as agricultural and livestock products; this type of industry elevates the Biochemical Oxygen Demand (BOD) levels.

There are many desalination plants in the RSA. Different methods are used (reverse osmosis, electrodialysis, thermal compression and vapor compression) but all cause pressures in the marine environment: by products of the desalination procedure (heated brine) disposed as discharges into the sea alter physical and chemical characteristics. Desalination discharges in volume account for 48 per cent of the total volume of effluents (Lattemann and Hopner, 2008).

The liquid waste only in Bahrain during 1996–1998 was about 1,500,000 tons/ year of tarry residues, followed by oil sludge (3,000–10,000 tons/year) and waste oil (12 tons/year). More information on industrial effluents are provided in the report by Al-Majed et al. (2000). Solid wastes originate from industrial and municipal sources. It is a substantial source of pollution. As an example, industrial solid waste in Bahrain amount to 35,000 tons per year whereas wastes from industrial activities in the eastern coast of Saudi Arabia are estimated to be about 1,200,000 tons per year (Al-Majed et al., 2000).

In addition to land-based sources of pollution, offshore installations of oil and gas located in the Gulf, contribute to the impacts on ecosystems' health. This is particularly serious in shallow waters and near ecologically sensitive areas. Transportation of huge quantities of oil, loading from terminals, operational discharges of oil tankers causing minor oil spills, result into accidents or pollution because of negligence from chronicle sources of oil pollution. Trace metals are not a threat for the environment as in most places their concentrations are not significant. Higher concentrations occur at a local scale near industrial centers.

4.10.3 Eutrophic conditions in the ROPME Area

Measurements of oxygen, chlα, ammonium, nitrates, phosphates and silicates are carried out within the framework of the national monitoring programs as well as during the Umitaka-Maru Cruises (1993–1994). It has been found that nutrient concentrations increased since 1975 when measurements started (Grasshoff, 1976). Sampling in Bahrain showed that the ranges of concentrations of nitrates, nitrites, ammonia and silicates were 0.38–0.77, 0.02–0.11, 0.04–16.60, 0.04–0.23 and 0.8–64.7 μgl^{-1}, respectively. The situation in Qatar was characterized by higher nutrient values: nitrates+nitrites (as one parameter) varied between 55.8–128.6 μgl^{-1}, whereas the ammonia values ranged between 67.4 and 127.2 μgl^{-1}. Phosphate concentrations were found between 1.67 and 1.76 μgl^{-1}. In the United Arab Emirates, nitrates were from 0.07–14.32 μgl^{-1} and nitrites 0.01–5.18 μgl^{-1}. It was concluded that sewage discharges in the area had increased nutrient concentrations (Shriadah and Al-Gais, 1999).

Primary productivity in the Gulf (Figure 4.18) is higher compared to the Red Sea but lower in comparison to the Arabian Sea (Sheppard et al., 2010). High productivity areas in the Gulf are estuaries and mudflats but generally speaking the Gulf system is nutrient limited. Chlα concentrations ranged between 0.2 and 0.86 mgm^{-3}, values not particularly high as opposed to chlα values in the Arabian Sea, exceeding 0.5 mgm^{-3} (Sheppard, 1993). Data from the oceanographic cruises during January 1993, December 1993 and December 1994 carried out by the R/V Umitaka-Maru, showed chlα concentrations ranging between 0.44 and 2.84 mgm^{-3}. Primary productivity measurements in the area showed photosynthetic rates ranging from 0.12 to 1.27 gCm^2d^{-1} (Al-Majed et al., 2000).

Sources of marine eutrophication in the RSA, apart from treated sewage from treatment plants and waste water of urban origin there are also effluents from industrial processes discharging methane/ammonia, oil refineries the livestock

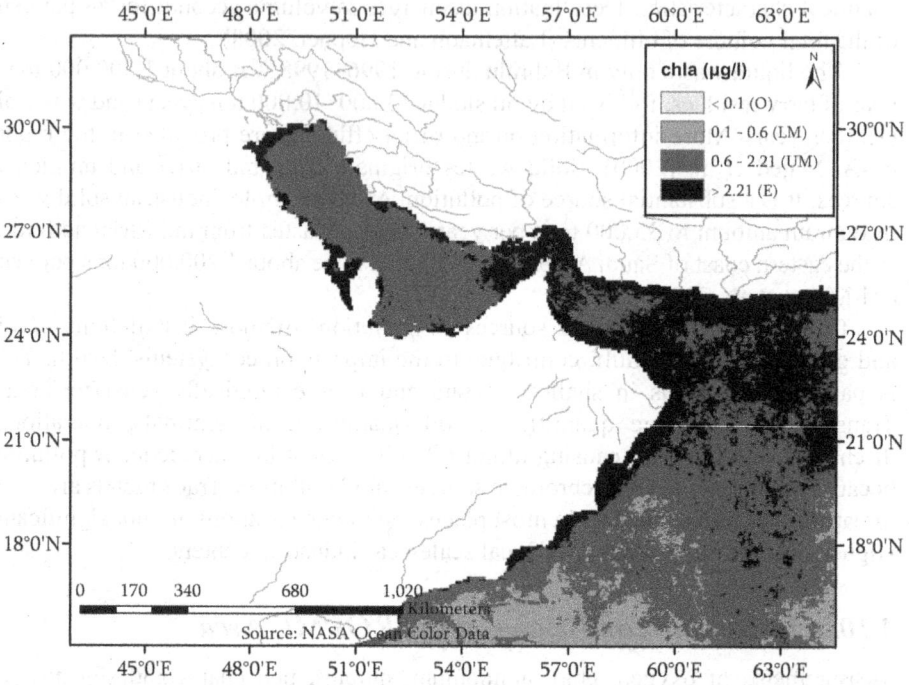

Figure 4.18 Surface chlorophyll distribution in the ROPME area during 2017. Eutrophication scaling according to Simboura et al. (2017-modified) scale.

industry and slaughter facilities. The major input of nutrients into the sea comes from rivers, receiving loads from sewage disposal sites and agriculture. Most of the nutrient loading remains in the shelf areas and only a small fraction of nutrients is driven into offshore marine areas. Nutrient discharges trigger algal blooms fairly often, even in the pelagic zone. These episodes form a threat to mariculture and coastal recreational activities. However, permanent eutrophic conditions are limited near urban areas. There is information on eutrophication on the northern coast of Bahrain where filamentous green algae form mats, in the Shatt Al-Arab area, the Kuwait Bay, the coastal area of Dhahran (Saudi Arabia), Abu Dhabi (United Arab Emirates) and Muscat (Oman). Red tides have appeared locally in Bahrain and Saudi Arabia. Although the situation from the point of view of eutrophication is reasonable, there are future trends due to the fast growing population and industrial developments.

4.10.4 *Policy effectiveness in the ROPME Sea Area*

The Kuwait Convention forms the legal framework for the ROPME States to protect the RSA using the Kuwait Action Plan. The Kuwait Convention states that "*the protection of the marine environment is considered as the first priority of the Action Plan, and it is intended that measures for marine and coastal environmental protection and development should lead to the promotion of human health and wellbeing as the ultimate goal of the Action Plan*". The Kuwait Convention is accompanied by four protocols, on combatting oil pollution and other harmful substances, marine pollution from the exploration/exploitation of the continental shelf, pollution from land-based sources and a protocol on control of transboundary pollution and disposal of harmful substances.

In spite of the cooperation among the eight Member States through the ROPME platform, the progress in the implementation of the Action Plan has not been very successful so far (UNEP, 2015). Doubts have been expressed on the effectiveness of the ROPME framework but there is a belief that it has got the potential to become more effective (Sale et al., 2011). According to Nadim et al. (2008), the main problem connected to reduced effectiveness is lack of ranking of crucial coastal issues.

In spite of the ongoing cooperation for about 40 years between the Member States, data and information exchange is limited (Sheppard et al., 2010). The information compiled so far would have had helped to understand processes and draw reliable assessment. The 2030 Agenda adopted by the Member States in September 2015 lists a number of goals: according to Goal 14.1, the Member States should try by the year 2015 to *"prevent and significantly reduce marine pollution of all kinds, in particular from land-based activities, including marine debris and nutrient pollution"*. Mitigation of nutrient pollution would ease anthropogenic pressures on the coral systems and other sensitive and vulnerable ecosystems so as to ensure ecosystms' integrity and functioning.

4.11 Eutrophication Status of the Red Sea and the Gulf of Aden

4.11.1 Physiography of the Red Sea and the Gulf of Aden

The Red Sea located between the Mediterranean Sea and the Indian Ocean, is a semi-enclosed body of water surrounded in the west by Egypt, Sudan, Eritrea and Djibouti, whereas in the east is bordered by Yemen and Saudi Arabia (Figure 4.19). The Red Sea is a narrow-shaped and elongated marine system extending from north to south over a distance of 2,000 km (PERSGA, 2006). The width is about 280 km on an average, whereas the maximum width 306 km, is located in the south. The width of the Strait that separates the Red Sea from the Gulf of Aden is about 26 km. The total area of the Red Sea is 438,000 km^2 (Figure 4.19). The northern part of the Red Sea is separated into two Gulfs: the Gulf of Suez and the Gulf of Aqaba. The Red Sea is a shallow basin at the northern and southern ends reaching a depth of about 2,000 m in the central area. The semi-enclosed character of the Red Sea restricts the exchange of water masses to the Indian Ocean; the residence time has been estimated around 200 years. Salinity values are high and there is a salinity gradient from south (38–39 psu) to north (42 psu); these northern water masses become heavier and sink below the thermocline, forming bottom water during the winter months. The temperature of the surface layer ranges on an annual basis between 22 and 32°C. The water budget in the Red Sea is negative because evaporation (2,000 mmy^{-1}) exceeds precipitation by far (10 mmy^{-1}), causing an inward flow from the Gulf of Aden through the Bab el Mandeb into the Red Sea (Sheppard et al., 1992). The renewal time of the top 200 m is about 6 years, much longer than the renewal of the whole water mass of

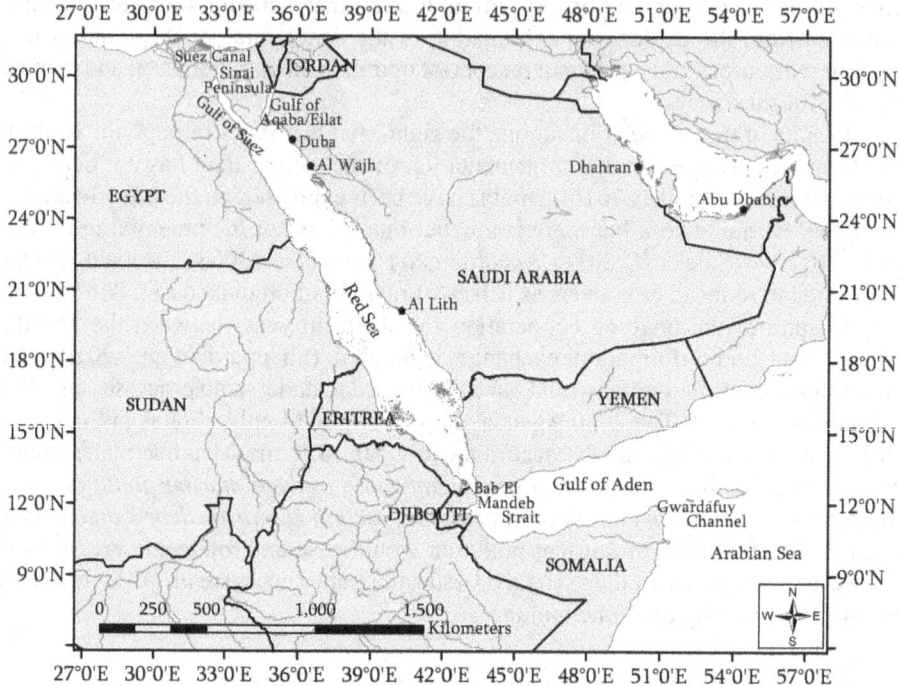

Figure 4.19 The Red Sea and the Gulf of Aden.

the Red Sea. There is a strong thermocline in the Red Sea throughout the year that isolates nutrient rich bottom waters from the photosynthetically active surface layer. Water currents in the Red Sea are wind-driven and influenced by the Indian Ocean Monsoons. The surface currents seem to move in an anticlockwise direction.

The Gulf of Aden is surrounded by Yemen to the North, the Arabian Sea and Gwardafuy Channel to the east, Somalia to the south and Djibouti to the west. It connects to the Indian Ocean through the Guardafui Channel. The length of the Gulf of Aden from west to east is 1,260 km whereas the distance from north to south is approximately 350 km. The surface area is about 410,000 km² and the average depth about 500 m. The wind regime in the Gulf of Aden is basically affected by the monsoons blowing in the Indian Ocean. Currents in the Gulf of Aden are wind-driven and stronger than in the Red Sea. The sea temperature in the Gulf of Aden varies between 15 and 28°C, depending on the appearance of monsoons and the season. Salinity values range from 35.5 psu on the eastern part of the Gulf to 37.3 psu in the Gulf's central area, whereas oxygen content fluctuates between 4.0 and 5.0 mgl^{-1}.

4.11.2 *Environmental pressures and sea conditions in the Red Sea and the Gulf of Aden*

Population growth and economic developments in coastal areas of the Red Sea has applied considerable pressures in the marine environment including marine ecosystems' destruction, toxic effects from discarding of chemicals and, at a local scale, nutrient discharges leading to eutrophication. A significant threat to the marine environment in the RSGA area is maritime transport. The number of ships passing through the Suez Canal is high and these numbers show an upward trend. Due to the semi-enclosed character of RSGA (the Red Sea and the Gulf of Aden), any oil accidents can have serious effects on the marine environment. There are also threats from refineries, liquefying natural gas terminals as well as from loading-unloading operations (PERSGA, 2006). Ballast water from tankers, containers and bulk carriers in the RSGA area are a potential risk not only for pollution effects but also for introducing alien species in the area.

Intensive coastal development in some areas of RSGA, instigate future pressures from land-based sources due to discharge of sewage and industrial wastewater, causing at a local scale, eutrophication problems. Besides sewage, pesticides are entering the marine environment as a runoff, being a point pollution problem. Elevated levels of heavy metals occur at a local scale in some areas. Desalination facilities are also connected with discharge of heated, concentrated brine and chemicals used during the desalination processes (Hoepner and Lattemann, 2003). The unique marine life (Sheppard et al., 1992) and the sub-tropical climate have favored the development of the tourist industry, especially in Egypt and Jordan followed, to an extent, by other RSGA countries. Tourism in Egypt is the third source of revenue after oil and remittances. Pressures on the coastal zone affects coastal ecosystems and increase eutrophication problems in the holiday resorts.

Marine ecosystems in the Red Sea and the Gulf of Aden area are dominated by coral reefs; it has been estimated that they cover more than 50 per cent of the total area. However, coral reefs are threatened by numerous human activities including discharges from urban and industrial centers, dredging, oil discharges and waste from shipping and scuba diving by tourists. Most of the coral reefs were impacted by coral bleaching in 1998, Abelson et al. (2005), especially in areas near Saudi Arabia and Yemen but now they are in the process of recovering. On the contrary, mangrove formations, seem to be heavily damaged: grazing by camels, cutting, burial by sand dunes and sewage are the most common causes of mangrove deterioration. The dominant mangrove species in the RSGA area are *Avicennia marina* and *Rhizophora mucronata*. Common species are also *Bruguiera gymnorhiza* and *Griops tagae* (Sheppard et al., 1992). The fact that 250 species of marine invertebrates and vertebrates live in mangrove systems, shows their ecological significance in the area.

So far 500 taxa of benthic algae have been identified in the RSGA area. They are mainly filamentous greens, red algae and small browns (Leliaert and Coppejans, 2004). In areas too turbid for coral growth, microalgae often cover hard substrates (Gladstone, 2000). Mistafa and Ali (2005) have reported 50 taxa of seaweeds (Chrysophyta, Pyrophyta and Rhodophyta), whereas Ormond and Banaimoon (1994) have reported 163 taxa of macroalgae in the Gulf of Aden. Seagrass beds also form significant ecosystems in the area. Eleven species of seagrass have been reported in RSGA area (Mistafa and Ali, 2005). In some areas, seagrasses form large beds. However, in the Gulf of Aden, especially along the coastline of Somalia, there are only few seagrass beds due to high pressures on the environment.

4.11.3 Eutrophic conditions in the Red Sea and the Gulf of Aden

The nutrient status in the Red Sea is of a typically oligotrophic sea (Figure 4.20). The only exceptions to the Red Sea oligotrophic characteristics is some localities off the Sinai Peninsula, urban coastal areas and the transition zone between the Red Sea and the Indian Ocean. Due to strong stratification, there is no mixing and upwellings known in the Red Sea area. Regeneration of nutrients resulting from winter plankton blooms enriches the surface water near Jeddah (Shaikh et al., 1986). Nutrient concentrations in the south Red Sea are higher due to inflowing water from the Gulf of Aden. Nutrient concentrations of phosphate and nitrate for the Red Sea area have been presented in the PERSGA (Programme for the Environment of the Red Sea and Gulf of Aden) Report on the state of the environment in the Red Sea (PERSGA, 2006).

High transparency characterizes the surface layer extending the eutrophic zone down to depths between 74 and 94 m in the northern part of the Red Sea (Stambler, 2005). High water transparency means that Primary Productivity (PP) in the Red Sea is low (0.21–0.50 gCm^{-2}d^{-1}) all over the year. Chlorophyll peaks in the northern part of the Red Sea can reach the value of 0.4 mg chlα.m^{-1} between February and May. Nutrients from terrestrial runoff are negligible due to unpredictable and irregular character of the rainfalls as well as because of their limited number.

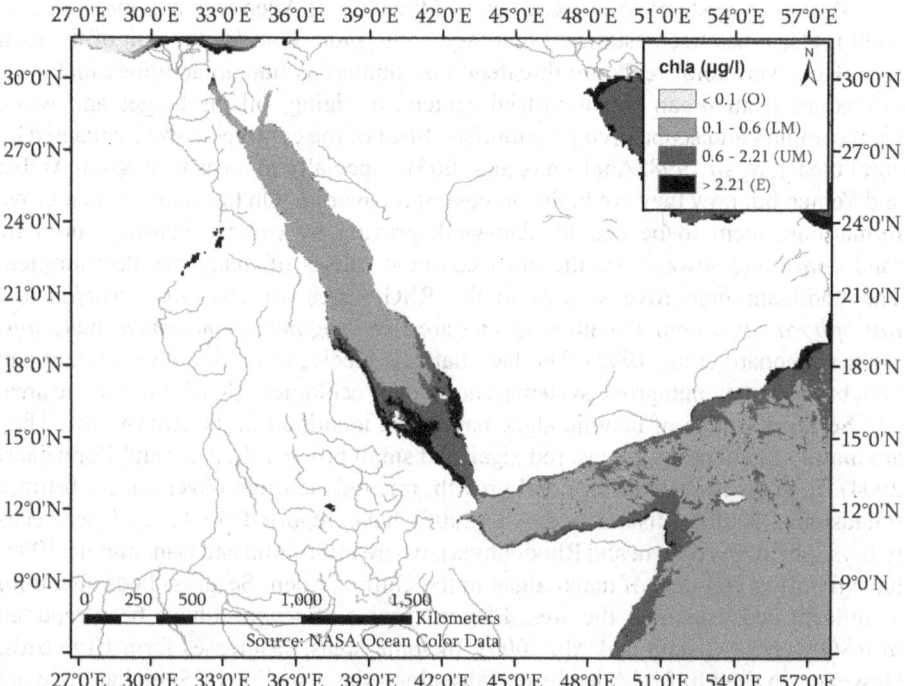

Figure 4.20 Surface chlorophyll distribution in the Red Sea and the Gulf of Aden area during 2017. Eutrophication scaling according to Simboura et al. (2005-modified) scale.

On the contrary, pelagic PP is higher (1.6 gCm^{-2}d^{-1}) in the Gulf of Aden, between May and September, along the Somalian coastline. It is not unusual, chlorophyll peaks to be as high as 3.5 to 5.5 mgm^{-3} in the surface layer during this period of time (Sheppard et al., 1992). Nutrient rich upwelling in the Gulf of Aden during the monsoon season, enriches the Red Sea and therefore enhances PP up to 19°N. Blue-green algal blooms dominated by the cyanobacterium *Oscillatoria eythreum* occur frequently in the Red Sea. They play a key role in supporting PP as they are nitrogen fixing organisms, providing ammonia and amino-acids, promoting succession of other phytoplankton groups.

It has already been mentioned that marine eutrophication is a local problem. Eutrophic trends have been reported by Ansari et al. (2015) in three selected sites (Haal, Sharmaa and Duba) along the northern coastal Red Sea region of Saudi Arabia. Although only one coastal site showed eutrophic trends, the very long Saudi coastal area in the Red Sea, characterized by developing human settlements (Al Lith, Al Wash, Duba, Haal, etc.) may be affected by the discharges of sewage; in that case, eutrophic spots are expected to develop. Experiments on the effects of eutrophic conditions on coral reproduction was studied in the Gulf of Eilat near an aquaculture unit (Loya et al., 2005). It was found that nutrients released from fish cages had a noticeable negative impact on the successful production of larvae of the coral *Stylophora pistillata*. Similar conclusions have also been drawn by Hall et al. (2018). The authors underlined the importance of maintaining oligotrophic conditions as a necessity to secure good health of the coral ecosystems. Eutrophication as a potential source of effects on coral communities has also been studied by Naumann et al. (2015). Experiments related to negative effects on corals under conditions of nutrient availability, have been performed by *in situ* experiments in the Central Red Sea (Jessen et al., 2012).

4.11.4 Policy effectiveness in the Red Sea and the Gulf of Aden

The "*Regional Convention for the Conservation of the Environment of the Red Sea and Gulf of Aden*" provided a foundation for the Intergovernmental Organization PERSGA (Regional Organization for the Conservation of the Environment of the Red Sea and Gulf of Aden); PERSGA is dedicated towards the conservation and management of the coastal and marine environment in the area. It has also carried out the *Strategic Action Plan* of the Red Sea and the Gulf of Aden. This institution played a key role in dealing with various threats due to population growth in the coastal areas, the economic growth and the maritime traffic that increased yearly at a remarkable rate.

It is obvious from the previous sections that in the RSGA area, there are many vulnerable and sensitive ecosystems that need attention and special measures to be taken. The main problem in the area is the effects of oil connected with transportation of crude oil through the Suez Canal and disposal of chemicals from industrial activities. The problem of eutrophication appears at a local scale mainly connected to coastal urban development (sewage discharges) and aquaculture facilities (nutrient release and organic pollution). It has already been confirmed by marine research in the Red Sea but also in other parts of the world that corals are particularly vulnerable to eutrophic conditions. It is therefore a priority to maintain oligotrophic conditions in the coral reef areas of the Red Sea and not simply to mitigate eutrophication.

The effectiveness of the RSGA Convention in eutrophication depends on the overall effectiveness of measures since there are no special actions focusing on eutrophication, although eutrophication is mentioned as the principal land-based pollution problem by almost all signatory members. Standard survey methods were adopted that helped the development of regional expert teams, capable of applying successfully these methods in each country. That was the initial step towards collaboration at international level. However, due to differences in priorities among the RSGA Signatory States, the Action Plans needed to be adapted to suit the needs and priorities of each country. They proceeded therefore, to the development of National Action Plans (NAPs) so as to adapt their monitoring surveys and at the same time to implement measures at national level. The NAPs formed clusters from national stakeholders, scientists, competent authorities and other relevant institutions. This administrative structure facilitated interstate collaboration in both, exchange of information and a common understanding of the problems and possible measures. The effectiveness of their policy was further enhanced by establishing within the RSGA framework a regional database in 2003 within the faculty of Marine Sciences, King Abdulaziz University in Jeddah. Training of personnel at various levels completed the setting of a common background at an interstate regime. The implementation of regional monitoring programmes has been a priority. This monitoring activity was focusing at the first stage on measurements on dissolved nutrient and physical–chemical variables in Egypt, Jordan and Saudi Arabia. Ecosystem conservation was rather site specific. There is an overall progress regarding environmental issues and the situation will become even better when the heterogeneity in research infrastructure and scientific personnel among the Member States of RSGA becomes smoother.

4.12 Eutrophication Status of the South-East Pacific

4.12.1 *Physiography of the South-Eastern Pacific*

Environmental Protection of the South-East Pacific area is under the aegis of the Lima Convention. The area of application of this Convention is the coastal zone and the sea area of the South-East Pacific within 200 miles' maritime area but even beyond that area up to a distance that high sea pollution can affect the Lima Convention area (Figure 4.21). The coastal zone of the Lima Convention area starts in the north in Panama and stretches along the Colombian, Ecuadorian, Peruvian and Chilean coasts down to Tierra del Fuego (Figure 4.21). The parties have agreed to take the necessary measures to protect the marine area under the Lima jurisdiction but mainly from land-based sources and coastal erosion.

The most remarkable physical phenomenon regarding circulation in the South Eastern Pacific Ocean is the Peruvian upwelling, which is having a major effect on the water circulation in the area Blink et al. (1983). The mechanism of the Peruvian upwelling is mainly a wind driven system; the mechanism is rather well understood (Brink et al., 1983): the Coriolis force induces a water mass transport, changing the direction by 90° to the left of the wind direction. The equator-ward wind pushes the

Figure 4.21 Lima Convention Region. It is the coastal zone from Panama to Chile down to Tierra del Fuego and 200 miles, off the Convention's coasts.

thin surface layer (10–30 m) to be directed offshore. This offshore movement results into a mass divergence near the coast and therefore, the outflowing mass has to be replaced by an upward movement of deeper water masses to counter balance the water mass deficit. Relatively recently there has been some concern about the Peru-Chile dynamics of the upwelling system under climate change. It has been proposed that the upwelling source will become shallower when sea temperature rises and fears have been expressed that this can seriously affect the biological productivity in the area (Oerder et al., 2015). Meanwhile, the poleward undercurrent, which flows below the surface layer of the continental shelf, supplies the upwelling water. Along the coastal waters of Panama and Colombia, seasonal cold-water upwelling occurs. In spite of the well-known significance of this phenomenon, not many physical and biological characteristics are known as yet.

Colombia together with Indonesia, Brazil and Mexico, are considered as the top countries in the world, regarding their biodiversity and for this reason they are known as "megadiverse countries" (Diaz and Acero, 2003). Although marine biodiversity in Colombian waters has not been studied as much as the biodiversity in the terrestrial environment, the fact that Colombian territorial waters are approximately 1,000,000 km² (on both sides though, the Caribbean and the Pacific Ocean), the state tends to be maritime by 50 per cent. The Pacific coast of Colombia is about 1,300 km; 25 per cent of the coastal zone is sandy beaches. Regarding marine ecosystems, coral reefs, although they occur on both seas of Colombia, only 2 per cent that is 20 km², are in the Pacific waters of the country. Seagrass beds do not occur in the Colombian Pacific waters. On the contrary, Colombian Pacific mangrove forests occupy about 3,000 km² and account for the 7.5 per cent of the total Colombian Pacific exploitable forest (Marrugo-Gonzalez et al., 2000). In Ecuador, an important ecoregion from the point of view of the marine environment is the La Region Insular, comprising of the Galapagos Islands, about 1,000 km west, offshore the coastline in the Pacific Ocean.

Fishing in Peruvian waters is a primary economic activity. Peruvian coastal waters are among the world's productive seas. There is an industrial scale of fishery, mainly focusing on pelagic fishes: anchovies account for 86 per cent of the catch but mackerel and squid also have a high commercial value. Peru is a leading producer of fishmeal and fish in spite of the fact that fish biomass fluctuates, often dramatically. The product output varies between 3 and 6 million tons. These big fluctuations seem to be linked to El Niño and La Nina (UN OECD, 2016).

4.12.2 Environmental pressures and sea conditions in the South Eastern Pacific

Information as it appears in international journals on environmental conditions along the SW Coast of Latin American water masses is scanty; the existing information in the international literature is limited mostly to local conditions. There is a lot of mining activity in the area and tailing waste seems to cause serious problems regarding heavy metal pollution. The second serious form of pollution that has already received attention is the plastics in the marine environment.

Peru is the biggest producer of gold in South America and the third copper producer worldwide after China and Chile. Peru is having mineral refineries that seriously pollute coastal waters. The problem of metal pollution is far more serious with artisanal mining because no measures for environmental protection are taken (UN OECD, 2016). The presence of metals, especially mercury in the Peruvian waters seems to be connected to the food chain. Mercury concentrations were measured in four piscivorus birds, endemic to the Peruvian Humboldt Current, mainly fed on the anchovetta. The results indicated that mercury was widespread among birds of the cold Humboldt Current; it must be emphasized that this current is of vital importance for the economy of the country including its value as a source of food (Gochfeld, 1980). There are also concerns in Peru regarding the accumulation of Cd and Pb in shellfish (Loaiza et al., 2015). Copper mining in Chile is an important activity and the mine tailing dumping, ending into the sea, causes impact on the coastal marine ecosystem of the Pacific. This impact has been studied as early as the 70s in the Chanaral Bay (Castilla and Nealler, 1978). Tailing waste consisting of suspended solids and chemical compounds has been reported during a 1976 survey when mass mortality had been caused in the subtidal and intertidal zones: starfishes, sea urchins, crabs, the shellfish "loco" (the shellfish *Concholepas concholepas*) as well as various other species were found dead.

An environmental assessment on sewage effluents has been conducted close to an industrial and urban area called Ventanas. The impact of the effluents containing nutrients and metals was studied on the invertebrate community, inhabiting a holdfast known as *Lessonia trabeculata*. Changes on species diversity were documented (Saez et al., 2012).

Floating Marine Debris (FMD) seems to be a common problem among the Lima Convention Signatory States. The States having recognized the multiple negative impacts of the long living plastics, are taking measures to eliminate the problem. In Chilian waters, the abundance of FMD during 2002–2005 was ranging between 1 and 250 items km^{-2}. It was also found out that 80 per cent of FMD consisted of expanded polystyrene, plastic bags and plastic fragments. Expanded polystyrene is used as a flotation device in mussel farm and plastic bags were coming from land-based sources (Hinojosa and Thiel, 2009). In Ecuador, it was found that 12 per cent of the stomach content of squids (*Dosidiscus gigas*) was plastic remains (Rosas-Lui, 2016). Due to the severity of the problem, in a "*Conferencia Internacional de Analisis de Ciclo de Vida en Latinoamerica*" (CILCA) that took place in Colombia (June 2017), the participants signed a declaration about improving the handling of plastic resources and develop new methods for mitigating the problem of marine litter (Sonnemann and Valdivia, 2017).

4.12.3 Eutrophic conditions in the South-East Pacific

The effects of urbanization of coastal areas in Chile were studied as a factor affecting the structure and function of coastal ecosystems due to eutrophication. This in turn has affected the presence of the shellfish (*Concholepas concholepas*) in Southern Chile. Elevated nutrient concentrations and higher phytoplankton biomass affect the size of the shellfish (small size), reducing their commercial value (Van Holt et al., 2012).

Frequent occurrence of HABs in the Southern Chile causes severe economic and social problems to the population in the area. Phytoplankton toxins can have a serious toxic impact on shellfish, fish, seabirds and sea mammals. HABs' presence has been documented since 1972, but meanwhile, the frequency of occurrence has increased alarmingly over the years. During 2016, a toxic algal bloom dominated by *Pseudochattonella* sp. killed 39 million of salmon, a number amounting to about 100 tons and causing a damage near US $800 million. Subsequent blooms were dominated by *Alexandrium catanella* in the Island of Chiloe and was characterized as the first bloom of that magnitude in the north (Foster and Hart, 2016). There are indications that *A. catanella* outbursts were connected to El Nino exceptional oceanic condition, but the main reason seems to be high nutrient concentrations. Chlorophyll distributions along the coastal area of the Lima Convention area is shown in Figure 4.22. This pattern indicates the presence of high nutrient concentrations that may be due to coastal pollution as well as to upwelling processes.

4.12.4 Policy effectiveness in the South-East Pacific Region

Not much information is widely available either on science or on policy of the effectiveness of the measures for the protection of the South Eastern marine environment, following Lima's Convention objectives and protocols. From the information mentioned above, it follows that the main environmental problem along the coastal environment from Panama to the Tierra del Fuego is metal pollution.

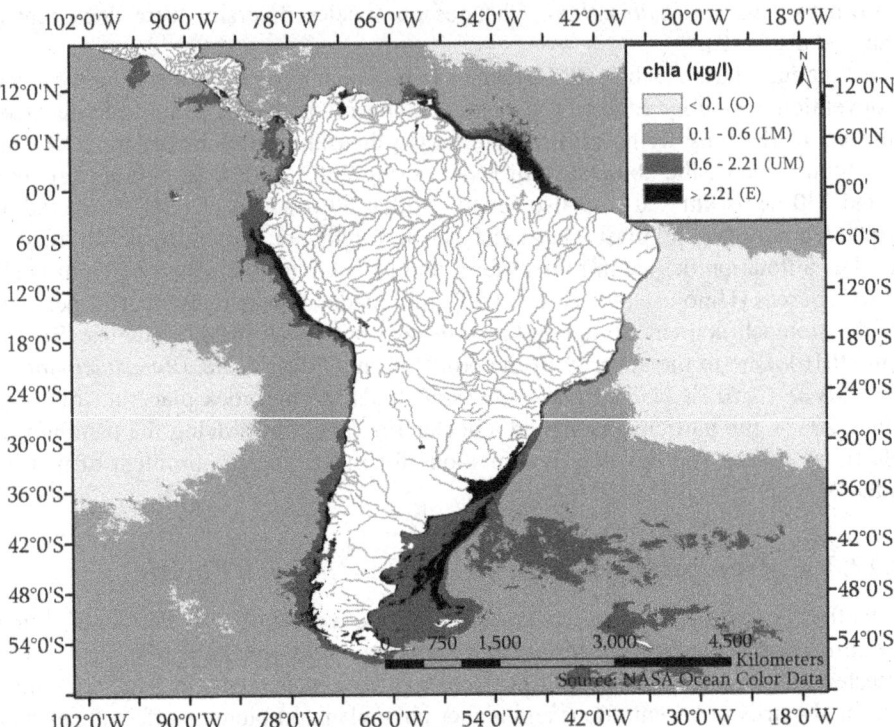

Figure 4.22 Surface chlorophyll distribution in the Lima Convention area during 2017. Eutrophication scaling according to Simboura et al. (2005-modified) scale.

Eutrophication is a problem of local interest and seems to cause a concern where it is connected with aquaculture and mass deaths from toxic algal blooms.

4.13 Eutrophication Status in the South Pacific

4.13.1 *Physiography of the South Pacific*

The Pacific Ocean is the largest Earth's Ocean. The area of the Pacific is 165,250,000 km^2 and covers 46 per cent of Earth's water surface (Figure 4.23). The Noumea Convention, also known as the South Pacific Regional Environment Programme (SPREP), is an organization with a total of 25 Member States (21 Pacific Island States and 4 Territories) that look after the quality of the marine environment.

The main circulation system in the South Pacific is the Southern Pacific Gyre bordered by the Equator to the north, the Atlantic Circumpolar Current to the south, the South American Continent to the east and Australia to the west. The two major components of the Southern Pacific Gyre are the South Equatorial Current, located about 15° N and the Peru/Chile current. There is a split near 18° S into the southward flow along the coasts of Australia and New Zealand and northward flow towards the Coral Sea and further north to Papua New Guinea. The circulation

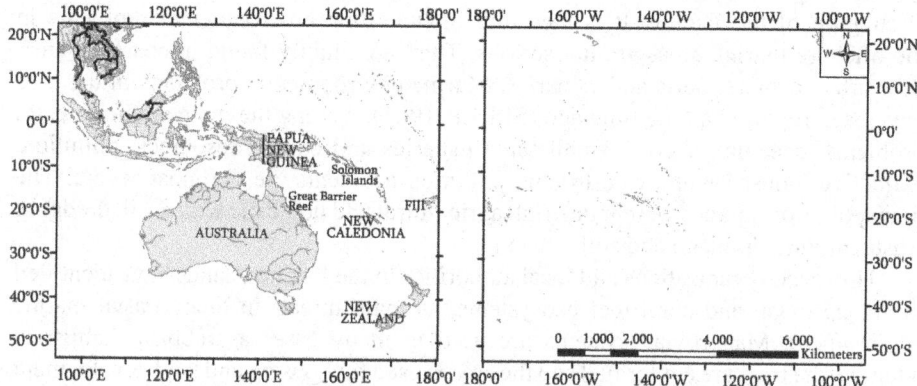

Figure 4.23 The South Pacific Region covered by the Noumea Convention.

pattern of the Southern Pacific Ocean is part of the Pacific Circulation. A more comprehensive description of the oceanographic conditions has been given by Tomczak and Godfrey (1994).

The most important ecotype of ecosystem in the area is the coral reef system. The biggest coral formation in the world is the Great Barrier Reef (GDR) on the northeastern Australian continental shelf and covers an area of about 350,000 km²; this long, narrow system stretches over 2,000 km along the coast and encompasses approximately 3,000 reefs. This huge marine ecosystem is the habitat for 350 species of hard corals, 1,500 species of fish, 4,000 species of mollusks and 240 species of seabirds (Brodie, 1999). Mangrove forests across the Pacific Region are also present in many parts of the area covered by the Noumea Convention: 45 mangrove species have been identified in the Papua New Guinea coastal systems, 33 in the Solomon Islands and 7 in Fiji. In many South Pacific States, the role of the mangrove forests has been recognized and therefore they are protected. They serve as nursery areas and also maintain fish populations of fish species important for the economy of the area (Brodie, 1999).

Estuarine systems in the area carry minerals and organic material into the marine environment. The water flow slows down in the estuaries so that its load of lighter particles settles. Marine plants make use of this nutrient rich background to grow, providing shelter to a large range of animals from worms and clams to larger animals including fishes. This type of estuarine system is common in the High Islands such as Papua New Guinea, New Caledonia and Fiji.

4.13.2 *Environmental pressures and sea conditions in the South Pacific*

The South Pacific Island States, signatories of the Noumea Convention, consist of islands scattered in the South Pacific Ocean. The land area of the small islands varies from 197 km² (American Samoa) to 29,785 km² (the Solomon Islands). The only exception is the Papua New Guinea (461,690 km²) and the States of Australia and New Zealand which are also signatory countries. In addition to limited population, the whole marine area is part of the Pacific Ocean and is open to the general circulation of the Pacific. It is therefore obvious that environmental problems in the offshore marine areas are not serious. They are mainly found in coastal waters near urban centers, ports and estuaries. Marine environmental problems in the area have been reported a long time ago (SPREP, 1981). Among the main environmental problems, oceanic fisheries, small scale fisheries and coastal resources, pollution, extractive industries and catastrophic pollution accidents are the most severe. The main pollution sources were industrial, agricultural and domestic waste, oil, dredging waste, mining drainage and soil erosion.

However, organizations and local authorities in the Pacific Islands have identified the mangrovian and coral reef ecosystems, as main threats in their coastal marine environment. Mangrovian systems are used in many cases as rubbish dumps; in addition, the trees are cut for fuel and the area is used for housing and land development in general. Some coastal constructions tend to interrupt the mixing between fresh and sea water; this creates an environment unsuitable for mangroves as most mangroves cannot survive in water as salty as the sea. They prefer brackish water where salt water is diluted by river water. As the mangrovian forests are a tourist attraction, tourists visiting these areas cause more problems to the natural system.

The coral systems are suffering from both natural enemies and human activities. There are many predators feeding on corals: the parrotfish scrapes the surface of the coral and fed on the polyps. The crown-of thorn sea star is another enemy, even more than the parrotfish. It has been reported that they destroy the coral reefs at a rate of 5 km a month (King, 2004) but the greatest threat to the corals comes from human activities. Coral blocks are used as building material; collection of corals for souvenirs is another threat. Harbor works and dredging, increase turbidity, attenuating therefore light penetration into the water column; suspended matter from dredging also affects various functions related to coral physiology. Corals are also destroyed by the use of explosives for fishing.

Although there are nutrient sources loading the marine environment from both waste and mining drainage, the South Pacific Seas are not affected by eutrophication. This is not surprising if we take into account that most of the Island States signatories in the South Pacific Region of Noumea Convention, are small islands with a total area of about 600,000 km², dispersed in an ocean area exceeding twenty million square kilometers that do not take into account marine areas beyond the 200 nautical miles from the land.

4.13.3 Eutrophic conditions in the South Pacific

The most common form of pollution restricted to coastal areas is sewage pollution as human waste is released into the sea. Although this is not a serious problem for the time being, as mentioned in the previous section, is bound to be a coastal eutrophication problem in future due to the rapid population growth. Sewage can be treated before disposal into the sea but sewage treatment is expensive and requires land for construction of ponds. Nutrient released into the marine environment in the area, support the growth of sea plants and phytoplankton blooms. Especially in lagoons, high levels of nitrates and phosphates have caused the overgrowth of the algal Sargassum, a species that caused a lot of destruction to the ecosystem by shading.

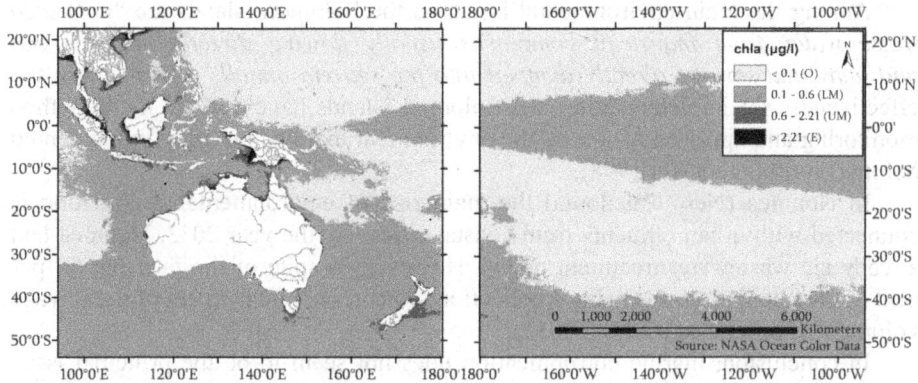

Figure 4.24 Surface chlorophyll distribution in the South Pacific during 2017. Eutrophication scaling according to Simboura et al. (2005-modified) scale.

Another problem relevant to microalgal growth and eutrophication is the *"cinguareta food poisoning syndrome"*. This syndrome is well known in Australia and generally the South Pacific Region. Humans consuming fish can suffer from gastrointestinal and neurological illness; in extreme cases deaths have been recorded as the result of respiratory failure (Hallegraeff, 2003). This syndrome is caused by excessive growth of some marine benthic dinoflagellates such as *Gambierdiscus toxicus, Ostreopsis siamensis* and *Coolia monafis*. These live in association with seaweeds and corals forming a film on their surface. These microalgae contain a chemical precursor which is converted into poisonous *"cinguatoxin"* in the livers of herbivorous fishes. They also live free on sediments and coral rubble. Although it has been a rare disease for centuries, is getting epidemic these days in the South Pacific Region. During the period 1960–1984, more than 24,000 cases have been reported in the region.

4.13.4 Policy effectiveness in the South Pacific

The Noumea Convention, adopted in 1986, functions within the framework of UNEP's Regional seas Programme. This Convention and its protocol entered into force on 22 August 1990. The Noumea Convention forms a legal system that provides protection and management of the Southern Pacific coastal and offshore environment, addressing pollution from all kind of sources, including land based sources (SPREP, 2010).

Although eutrophication is not explicitly mentioned (SPREP, 2010), a concern has been expressed by Australia regarding the causes of biodiversity decline: marine pollution as well as impacts from land-based sources, urban and agricultural pollution and point source emissions are included. Sewage and nutrients are also mentioned. Australia has also implemented a "Reef Water Quality Protection Plan" (Reef Plan) to stop water quality decline that enters the Great Barrier Reef. The Reef Plan provides actions to minimize, *inter alia*, nutrient pollution which is the main driver for marine eutrophication.

Among the main environmental issues in the Solomon Islands are *"untreated waste water from industrial companies usually flowing directly into the sea and untreated sewage directly dumped into the sea via outfalls"*. To increase the effectiveness for implementation, the Solomon Islands have decided to strengthen monitoring and apply an action on the environment that had not been implemented before (Environment Act 1998).

In Noumea (New Caledonia) the main risk of environmental degradation is connected with urban effluents from coastal cities. By the year 2013, Noumea had already sic wastewater treatment plants. However, in spite of the fact that 90 per cent of the population was using a collection system, only 30 per cent of the sewage volume was processed (OCTA, 2015).

In conclusion, marine eutrophication does not seem to be the principal issue in the South Pacific Area. It is locally connected with urban activities. No general conclusion can be drawn about eutrophication in the Region due to a number of reasons: (a) different sizes of the South Pacific territories (b) a different degree in the development of the States (c) lack of the data from some small States and (d) focusing on economic priorities. The main concern of the South Pacific States is biodiversity conservation, especially in the coral reef systems. Abatement of eutrophication is a prerequisite for "clean water" in the coral areas. Although they do not seem to face serious problems on marine eutrophication, it is not possible to assess, based on the published information, effectiveness of measures regarding the wider South Pacific Region, neither effectiveness in collaboration among the Member States of the Noumea Convention.

4.14 Eutrophication Status of the Western Indian Ocean

4.14.1 *Physiography of the Western Indian Ocean*

The Indian Ocean is the third biggest Ocean in the world; the area of the Indian Ocean is about 70,000,000 km^2 and the average depth about 4,000 m. However, the Nairobi Convention is limited towards the protection of the region of the Western Indian Ocean (WIO). The Convention area (Figure 4.25) is bordered on the west by Somalia, Kenya, Mozambique, South Africa and in the north by the Arabian Peninsula. Some of the Member States are Islands: the Republic of Seychelles, the Federal Islamic Republic of the Comoros, the Republic of Madagascar, the Republic of Mauritius and the Island of Reunion (France).

There is a large boundary current in the Western Area of the Indian Ocean that reverses twice a year, responding to wind reversals: the north-east winter monsoon and the south-west summer monsoon (Wyrtki, 1973). This current is as strong as the Gulf Stream and shows the highest variability in the world ocean circulation. Circulation at intermediate depths (1,000 m) is dominated by the anticyclonic gyre in the southern part of the Indian Ocean. Circulation in the Arabian Sea is weak. Lower than 3,500 m, the circulation is strongly influenced by the bottom topography. The oceanographic processes mentioned above influence the ecology of the region as they control nutrient availability of the surface layer (0–200 m), therefore inducing primary productivity, phytoplankton abundance and phytoplankton distribution,

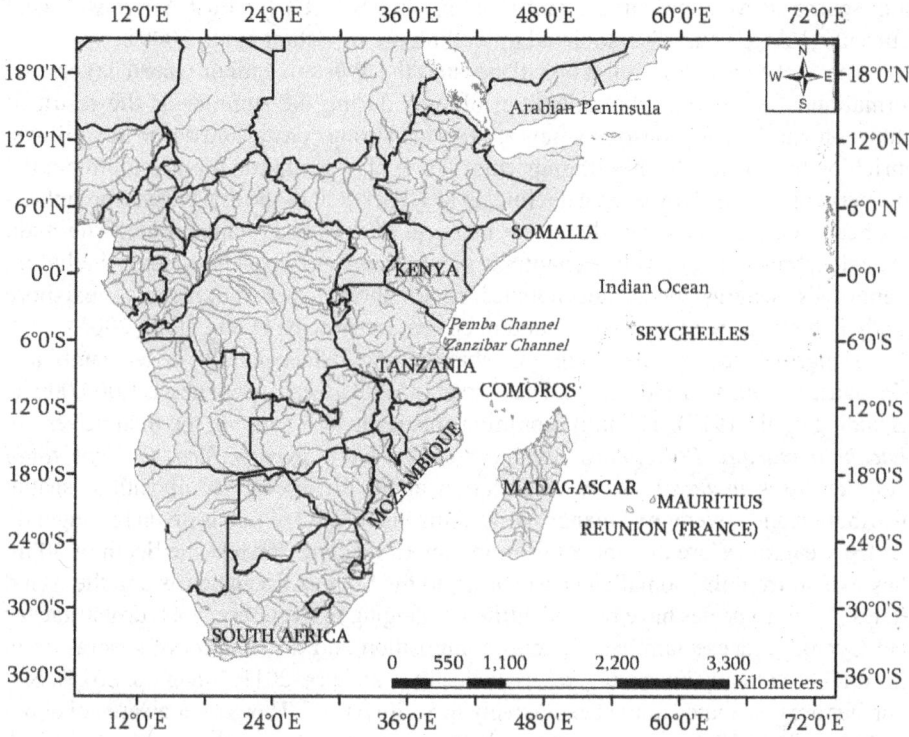

Figure 4.25 The Western Indian Ocean Region.

eventually affecting fisheries. These processes increase biodiversity in the WIO and also affect ecosystems' health. There are coastal currents, for example the East African Coastal Current (EACC) that are key mechanisms in larval dispersion. Currents are important on inshore areas, reefs and inlets where local currents are mixed with freshwater inputs, providing the major source of foods. Seeds of mangroves are also dependent on currents for their dispersion. Many animal species use the currents to navigate and reach feeding and reproduction sites. Currents also form a mechanism for the oxygenation of water masses, a prerequisite for benthic life (UNEP–Nairobi Convention and WIOMASA, 2015). Climatic conditions in the area are characterized by high humidity of tropical type, mean monthly temperatures, higher than 20°C and rainfall height more than 700 mm per year. The western coasts of the Indian Ocean suffer from strong cyclones (Couper, 1983). They usually occur shortly before and after the monsoon rains (Verlaan, 1991).

The oceanographic regime of the Western Indian Ocean is influenced by four factors: (a) the Asian Continent in the north (b) Madagascar (c) the Madagascar plateau and (d) the interactions of these factors with the boundary currents (equatorial and western) of the Ocean Basin (Obura et al., 2012). The main processes regarding oceanic circulation in the WIO are: (a) the South Equatorial Current (SEC) (b) the South Equatorial Countercurrent (SECC) (c) the Somali Current (SC): the striking seasonal reversal of the SC is a unique feature of the Indian Ocean (d) the Northeast and Southeast Madagascar Current (NEMC and SEMC) (e) East African Coastal Currents (EACC) and (f) associated upwelling areas (Schott et al., 2009).

The fact that among all tropical oceans, the Western Indian Ocean favors the formation of enormous phytoplankton blooms during the summer, is the result of monsoon winds. The monsoon winds are forcing strong coastal and ocean upwellings, enriching the surface layer with nutrients. These phytoplankton blooms maintain the marine ecosystems. However, a decline up to 20 per cent of the marine phytoplankton has been declined, possibly due to an increase of the ocean temperature. The main coastal habitats in the WIO, encompass estuarine and coastal systems including mangroves, seagrass beds, salt marshes, rocky shores, reefs, coral reefs, nearshore sandy habitats as well as offshore and deep sea ecosystems (Sling et al., 2005).

Mangrove ecosystems occur in estuaries and deltas mainly in Mozambique, Madagascar, Tanzania and Kenya. They cover in WIO a total area of about 1,000,000 ha (Spalding et al., 1977). The most common species in WIO among the mangroves are *Avicennia marina, Rhizophora mucronata, Brughiera gymnorrhiza, Ceriops tagal* and *Ceriops somalensis*. More information about the species composition, spatial distribution and coverage of mangrove systems in the WIO region is given by Lugendo (2015). Seagrasses are distributed from the intertidal zone down to the depth of 40 m; they extend from the Somalian coasts down to the South African coasts and the island States. Twelve species have been identified belonging to Zosteraceae, Hydroharitaceae and Cymodoceaceae families. Species composition and distribution of seagrasses in the WIO has already been studied satisfactorily (Bandeira, 2011; Lugendo, 2015). Salt marshes are distributed almost exclusively in South Africa. They cover an area of about 2,500 ha of the WIO region in the South African State. Their habitats are subtropical, intertidal and in floodplain areas. There is a decline from south to the north, following a temperature grade from temperate to subtropical regions. Their species diversity is limited, dominated by the species *Spartina maritima, Sarcocronia tegenaria* and *Salicornia meyeriana* (Lugendo, 2015).

The best known marine habitats in the WIO Region are coral reefs; they extend down to 50 m depth, provided that there is adequate transparency that can maintain the photosynthetic activities of the zooxanthelaceae. Coral formations are classified into four groups: fringing reefs (around the East African coast and the Islands), barrier reefs (developed in Madagascar), atolls (like the ones found tin Seychelles), coral banks, coastal and oceanic, occurring in many WIO places (Griffiths, 2005; Wafar et al., 2011). The shelves of the WIO continental margins form the transitional zone between the steep continental slope and the attached coastal plain. The shelves are also a transition zone for sediments transported from coastal areas to the deep ocean basins. They form habitats rich in invertebrates including macrofauna such as amphipods, polychaetes and gastropods, meiofauna such as rotifers, harpacticoids, nematodes and copepods (Fennesy and Green, 2015). There are also diatoms, flagellates and cyanobacteria forming a layer on shelf sediments in the euphotic zone which contribute to continental shelf productivity (Gattuso et al., 2006). The deep sea habitats in the WIO have been poorly known, as the deep sea research so far in the area is limited. Nevertheless, their connection with marine eutrophication processes is very weak, if it exists at all.

4.14.2 Environmental pressures and sea conditions in the Western Indian Ocean

Population characteristics in the WIO Region vary among the Member States but they have a common characteristic: high growth rates. Tanzania's population growth rate is relatively high in the area (data of the year 2014), followed by Madagascar and Maozambique. These growth rates are attributed to natural growth and in some states like Kenya, Tanzania, Mozambique and Madagascar are as high as 2.0 per cent per year. The total population of WIO States was estimated to be about 212 million in 2014, whereas there are estimates that this population will double (412.64 million) by the year 2050. Population densities are quite high in the Seychelles, Comoros and Mauritius (Mahongo and Mwaipopo, 2015). There is rural to urban migration as people tend to move to coastal urbanized areas with economic activities. It has been reported by Francis and Torell (2004) that a third of the Mozambicans and a quarter of the Tanzanians live by the coast. With the exception of Madagascar, which is a big island, there is no distinction in the other Island States between populations inhabiting the coastal zone and inland populations due to their small sizes. Apart from Kenya, about 50 per cent of the population of the WIO African countries live by the coast.

The economies in the WIO Region are characterized by different growth rates amongst them, varying between 4 and 7 per cent. Economic activities include transport, construction, agriculture, mariculture and manufacturing. Tourism and services in general are also among the activities, especially in some areas such as Reunion's economy where 82 per cent of the revenue comes from tourism. Tourism is a principal source of income in the WIO area, especially in small Island States, which is practically the base for their economy. However, tourist growth is a serious pressure and there is a growing environmental concern (Gossling, 2006). Pressures for property development in coastal areas, sewage pollution from hotels, habitat degradation especially coral formations, overfishing, coral sand mining and oil spills are the most common causes of marine environmental degradation. Inadequate marine monitoring and inadequate environmental impact studies cause additional difficulties in taking or implementing environmental protection measures.

The development of urban areas along the coastal zone, places a number of environmental issues: high demand on coastal, natural sources (extractive and non-extractive). Conversion of coastal areas for land development affects coastal systems, their diversity and their sustainability. Effluents, mainly of sewage and industrial origin have a broader effect on the quality of the marine environment. Microbiological contamination and solid waste have to be added to the sources of pollution of urban origin. Impact on the marine environment from urban WIO areas has been given by Cellier and Ntombela (2015).

All these activities mentioned above place many threats on the marine ecosystems. As coral reefs are highly valued for their ecosystem services in the WIO Region, they are subject to significant exploration. In addition to exploitation, urban pollution, coastal development, tourism and terrigenous sedimentation are the main sources of coral deterioration.

Coral diseases show an upward trend following the human footprint in the WIO areas; outbreaks of the thorn starfish (*Acathaster planci*), introduction of alien species and the damaging mussel *Crassostrea gigas* are the most common natural enemies of the coral formations (Obura, 2015). Threats to mangroves mainly come from overharvesting. This can be done for firewood, timber and charcoal, land reclamation for development, salt production, agriculture, tourism and construction of hotels and residential areas (Lugendo, 2015). Saltmarshes are suffering from human and natural factors. Water abstraction and human construction in the river mouths are the main problems for the declining of the salt marshes. Beaches and nearshore areas, in addition to the impact from the climate change, suffer physical disturbances such as runoff from land, habitat modifications and pollution from land-based sources. A summary of the Drivers–Pressures–Status–Response (DSPIR) is given for the WIO nearshore environment by Maina (2015).

4.14.3 *Eutrophic conditions in the Western Indian Ocean*

There are two large marine ecosystems in the WIO Region: (a) the Somali Current Large Marine Ecosystem (SCLME) that covers an area from the northern part of Madagascar and the Comoro Island to the Horn of Africa in the north. The SCLME is bordered by Tanzania, Kenya and Somalia (b) the Agulhas Current Large Marine Ecosystem (ACLME). It covers the southwestern Indian Ocean, including the continental shelves of Mozambique, the archipelagos of the Comoros, the Seychelles, Mauritius, La Reunion–France and Madagascar at the northeastern end (Heineman et al., 2008).

Ocean currents are important as they influence UNEP-Nairobi Convention and WIOSMA: The Regional State of the Coast Report: Western Indian Ocean. UNEP and WIOSMA, Nairobi, Kenya. Among the nutrients in the WIO area, nitrogen and phosphorus seem to be the limiting nutrients, whereas silicate does not seem to be limiting in the region (Kyewalyanka et al., 2007). There are significant spatial variations in the nutrient distribution in the WIO area but vertical distributions have a common profile: low nutrient concentrations in the euphotic layer due to exhaustion from nutrient uptake by phytoplankton. Nutrient concentrations in various parts of the WIO Region are given by Kyewalyana (2015).

Both the LME areas are moderately productive: average PP values ranging between 150 and 300 $gC^{-2}y^{-1}$. However, some coastal areas are very productive. There is also seasonal variation in PP: the WIO waters during the summer are characterized by low current concentrations, warm sea temperatures, low phytoplankton biomass and relatively low primary production (Karl et al., 2011). In more recent work (Barlow et al., 2011) it has been found that PP in Pemba and Zanzibar Channels ranges from 0.79–1.89 $gC^{-2}d^{-1}$. Most of the daily PP values in the WIO Region are within the range of 0.5 and 2.0 $gC^{-2}d^{-1}$. More information on PP has been published over many areas about the WIO Region (Kyewalyana et al., 2007; Ryther et al., 1966; Ryther et al., 1971).

Indications on eutrophication at a large scale in the WIO Region has not been confirmed so far (Figure 4.26). It only occurs locally in some coastal areas due to input of wastewater (UNEP–Nairobi Convention Secretariat, 2009). However, hot spots of harmful or nuisance algal blooms (HABs) have been observed. In regional surveys conducted between 1998 and 2000, more than 60 potential HAB species have been identified in the WIO area. Although Mauritius had the biggest HABs impact, HAB events have also been recorded in Kenya, Somalia, South Africa and Tanzania (ASCLME/SWIOFP, 2012). The blooms affected marine environmental quality, tourism and the local communities.

Some countries in the WIO area try to combat eutrophication and consequently the frequent occurrence of HABs, by reducing the nutrient load of wastewaters through biological sewage treatment. There are some estimates regarding municipal wastewater in some WIO States: wastewater disposal in Comoros is 168,000 m^3d^{-1}, whereas in South Africa is about 255,000 m^3d^{-1} (UNEP–Nairobi Convention Secreteriat and WIOMSA, 2009b). Nutrient inputs also originate from aquaculture activities. Aquaculture units contribute considerable amounts of nutrients and they are therefore a potential risk for harmful bloom formations. The solution to the problem can be either sustainable aquaculture policy or an integrated, three component mariculture system including finfish, shellfish and seaweeds (Mmochi and Mwandya, 2003). In any case more monitoring work is needed on nutrient availability and distribution as well as studies on the trends of primary productivity and its environmental social and economic impact.

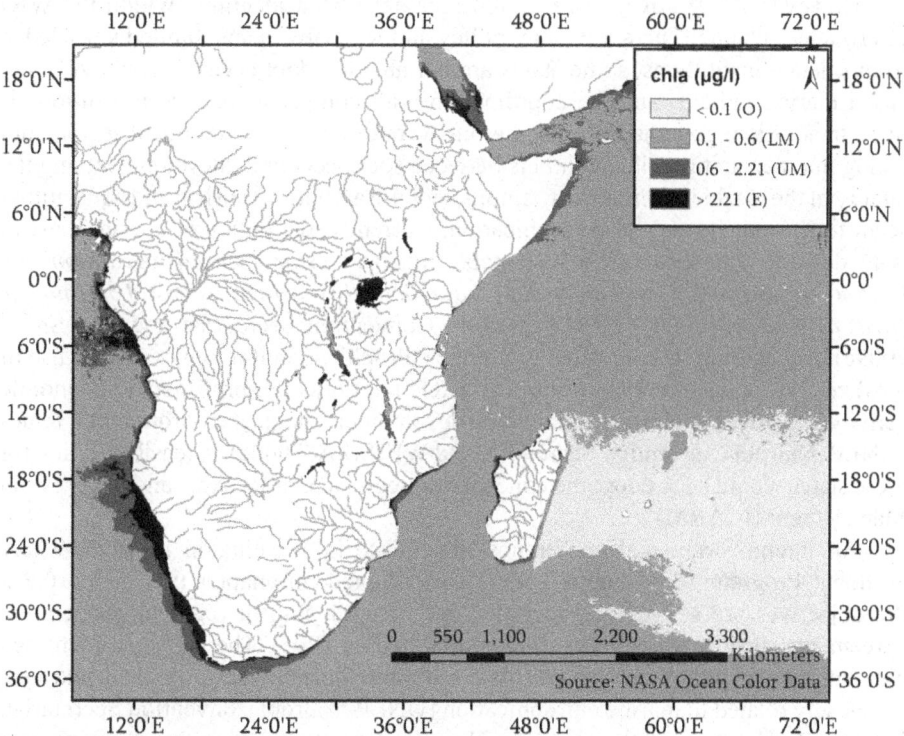

Figure 4.26 Surface chlorophyll distribution in the Western Indian Ocean during 2017. Eutrophication scaling according to Simboura et al. (2005-modified) scale.

4.14.4 *Policy effectiveness in the Western Indian Ocean*

The Western Indian Ocean Region, is also referred to Eastern and Southern Africa. The WIO Region has been globally characterized as an area with extreme biological and ecological richness, providing high socio-economic value. Due to this uniqueness and at the same time the pressures and threats on the quality of the marine environment, the African States of the West Indian Ocean signed in 1985 the Nairobi Convention (Momanyi, 2015). This Convention forms a legal system for the management, protection and development of the marine environment in the Eastern and Southern African Region. The Nairobi Convention is the main framework among many institutions of governance in the WIO Region that include national, regional and global institutions as well as regulating and policy frameworks. Among the regional institutions relevant to marine environmental quality, are the African Ministerial Conference on the Environment (AMCEN) and the New Partnership for Africa's Development (NEPAD). AMCEN was established in 1985 and is leading the process for the development of the Action Plan endorsed in NEPAD. The NEPAD Environment Action Plan has defined priority sectors: combatting land degradation, invasive species, wetlands, marine and coastal resources and climate change (UNEP–Nairobi Convention Secreteriat and WIOMSA, 2009b). It is obvious from the above mentioned objectives that there is no specific mentioning of nutrients, phytoplankton or eutrophication.

However, many governance weaknesses have been identified within the WIO governance scheme. There are many policy and legislative gaps, limited knowledge from science institutions, some states are not able to adopt decentralized evidence-based marine policies and the coordination with competent science institutions is poor. In addition, governance processes have concentrated on individual sectors, not taking into account possible conflicts between socio-economic development and the quality of the marine environment. Among sectoral activities, tourism and aquaculture seem to be connected with eutrophication. Currently, the policy in coastal tourism is to develop this sector in a sustainable manner "*using focused legislation and institutions and policy instruments that are sufficiently sensitive to coastal and marine environmental concerns*" (UNEP–Nairobi Convention Secretariat and WIOSMA, 2009c). Aquaculture is connected to marine eutrophication through excessive use of fertilizers. As aquaculture in most of the WIO States is a high-profile socio-economic activity, the governments prioritizing the development, neglect environmental issues (UNEP–Nairobi Convention Secretariat and WIOSMA, 2009c). A feasible policy for these states would be to adopt the concept of Integrated Coastal Area and River Basin Management (ICARM).

An intergovernmental conference took place in Washington in 1995, where a Global Programme of Action (GPA) was adopted. In chapter two of the GPA, the objectives are set out and include "the strengthening of regional cooperation agreements such as the Nairobi Convention" (GPA, 1995). The GPA had identified nine sources of pollution; two of those, i.e., municipal wastewaters and nutrients are closely related to marine eutrophication (UNEP–Nairobi Convention Secretariat, 2009a). Conferences of the parties within the framework of regional agreements enhance institutional cooperation and coordination. The fact that there is a satisfactory degree of similarity of "*interests, norms, perceptions and values at the regional level*" also strengthens the effectiveness of the cooperation among the WIO Member States.

During the Eighth Conference of the Parties to the Nairobi Convention in June 2015, it was decided to "*encourage collaboration and communication between contracting parties and civil society, private sector, non-governmental organizations, local governments and municipal authorities*" (Decisions for COP8, 2015). The Decision COP8 came to a resolution to establish additional partnerships, *inter alia*, with the Indian Ocean Commission and United Nations Agencies: The Southwest Indian Fisheries Commission on Sustainable Fisheries Management and the Convention on Biological Diversity, among them, are relevant to marine eutrophication.

In conclusion, WIO States have not characterized eutrophication as a major issue. They face it at a local scale and therefore, there is no explicit mentioning in the Nairobi Convention, in spite of the HAB events that seem to be a nuisance to tourism and mariculture activities.

The Epilogue: Eutrophication and Marine Policy

Different activities in the marine environment, have shown an upward trend over the last decades in both, variety and intensity, creating conflicts and environmental issues. These problems have brought into the surface the issue of sustainability or sustainable development. It is obvious that eutrophication is only a component in the economic, social and environmental system of benefits and shortcomings from the exploitation of the marine environment. Mitigation of the seriousness of environmental impacts, assumes action towards two interacting directions: marine spatial planning and environmental protection schemes. Marine Spatial Planning (MSP) is understood "*as the strategic placement of human activities at the sea as to achieve the regulation, management and protection of the marine environment in such a way as to mitigate if not minimize conflicts and negative effects on the marine ecosystem and to increase synergy*" (Kyvelou and Pothitaki, 2017). The main objectives of MSP are: (a) investment evaluation and future prospects in marine sectoral activities (b) conflict prevention or reduction between different activities through integrated planning and (c) formulation and implementation of the "ecosystem based approach". The last objective forms an interlink between resource development and environmental protection. The main marine activities globally recognized for their environmentally negative effects are illegal, unreported and unregulated fishing, trawling of high seas seamounts, geo-engineering, drilling for oil and gas. At the same time land activities and their disposals find their way to the oceans: sewage treated or untreated, garbage, fertilizers, pesticides, industrial chemicals, marine debris, introduction of alien species and hazardous nuclear waste are among the most important threats for the marine environment (Van Dyke and Broder, 2017).

The first legal framework for the protection of the marine environment was in 1982 the UN Convention on the Law of the Sea. In particular, Article 209(2) requires that state Parties should "*adopt laws and regulations to prevent, reduce and control pollution of the marine environment from activities in the area undertaken by vessels, installations, structures and other devices flying their flag or of their registry or operating under their authority*". But the most concrete initiative of UN was the launching in 1972 of the United Nations Environmental Programme (UNEP) which has become up to date the leading force in developing regional regimes. UNEP's objective was to address marine environmental affairs through Regional Seas Programmes (RSA); within the framework of the RSA, Regional Action Plans were elaborated. Each Regional Action Plan is composed of five components: (a) environmental assessment (b) environmental management (c) legal requirements (d) institutional arrangements and (e) funding.

As every marine geographical region is covered by a Convention under UNEP's aegis, it is surrounded by several member states and most of the pollution problems are of transboundary nature, including eutrophication. There is a need for cooperation among the signatory countries. According to Principle 24 of the 1972 Stockholm Declaration, States should cooperate *"through multilateral or bilateral arrangements or other appropriate means is essential to effectively control, prevent reduce or eliminate adverse environmental effects resulting from activities conducted in all spheres in such a way that due account is taken of the sovereignty and interaction of all states"*. In addition, emphasis on land-based sources of pollution which are closely connected with eutrophication, has been placed in Agenda 21: *"Land-Based Sources contribute 70 per cent of marine pollution, but that there is currently no global scheme to address marine pollution from Land-Based Sources"*. There are many sources of pollution from land, some of them difficult to identify and even more difficult to quantify, especially non-point sources and run-off sources and most of them are connected to eutrophication such as sewage, mining, agriculture and animal husbandry. Although the scientific estate has developed methods of assessment and norms to govern the marine environment, relatively recent issues that is climate change and technological advances, reset the existing management practices and are challenging policy makers to decide on how to cope with sustainability having these two components added to the ocean policy system (Van Dyke and Broder, 2017).

A necessary step in the whole procedure is the assessment of the effectiveness. By effectiveness we mean the Outcome Evaluation and Organizational Process Evaluation (Jacobson, 1995). The endpoints of evaluation include living resource population increases, improvement of marine environmental quality and decrease of pollutant discharges. The framework briefly mentioned above, is the domain. in which eutrophication problems should be identified, assessed and dealt with. Eutrophication evaluation presented in Chapter Four, has shown that the priority of eutrophication problems differs in the different regions examined. In some cases, it is indirectly involved with other environmental problems whereas in other cases is a problem of local interest. Although marine environmental quality studies can be focused on marine eutrophication, management and policy aspects, eutrophication will be only a component built-in, in the ocean policy framework.

Bibliography

Abelson, A., Olinky, R. and Gaines, S. (2005). Coral recruitment to the reef of Eilat, Red Sea: temporal and spatial variation and possible effects of anthropogenic disturbances. Marine Pollution Bulletin, 50: 576–582.

Ackefors, H. and Enell, M. (1990). Discharge of nutrients from Swedish fish farming to adjacent sea areas. AMBIO, 19: 28–35.

Adams, J.B. (2016). Distribution and status of Zostera capensis in South African estuaries. A review. South African Journal of Botany, 107: 63–73.

Agardy, T. (2010). Ocean Zoning: Making Marine Management More Effective. Earthscan. London. 220pp.

Agenda 21. (1992). Proceedings of United Nations Conference on Environment and Development, Brazil, Rio de Janeiro, UN.

Aghai Diba, B. (2003). Pollution in the Caspian Sea. Payvand Iran News. Available at: http://www.payvand.com/news/02/jul/1073.html.

Ajuzie, C.C. and Houvenaghel, T. (2009). Preliminary survey of potentially harmful dinoflagellates in Nigeria's coastal waters. Fottea, 9: 107–120.

Aladin, N.V. and Plotnikov, I.S. (2004). Hydrobiology of the Caspian Sea. In: Nihould, J.C.J., Zavialov, P.O. and Micklin, P.P. (Eds.). Dying and Dead Seas Climatic Versus Anthropic Causes, NATO Science Series: IV: Earth Environmental Sciences, Vol. 36. Springer. 437.

Albert, S., O'Neil, J.M., Udy, J.W. et al. (2005). Blooms of cyanobacterium Lyngbya majuscule in coastal Queensland, Australia: disparate sites, common factors. Marine Pollution Bulletin, 51: 428–437.

Alcaraz, M., Estrada, M. and Marrase, C. (1989). Interaction between turbulence and zooplankton in laboratory microcosms. Proc. 21st. E.M.B.S. Gdansk. pp. 191–204.

Alexander, T.J., Von lanthen, P. and Seehausen, O. (2016). Does eutrophication driven evolution change aquatic ecosystems. Philosophical Transactions of the Royal Society B372, 20160041. Available at: http://dx.doi.org/10.1098/rsth.2016.0041.

Al-Majed, N., Mohammadi, H. and Al-Ghadban, A. (2000). Regional Report on the State of the Marine Environment. ROPME/GC-10/001/1. Revised by Al-Awadi, A., Regional Organization for the Protection of the Marine Environment.

Almasri, M.N. and Kaluarachchi, J.J. (2005). Multi-criteria decision analysis for the optimal management of nitrate contamination of aquifers. Journal of Environmental Management, 74: 365–381.

Aps, R. and Lassen, H. (2010). Recovery of depleted Baltic Sea fish stocks: a review. ICES Journal of Marine Science, 67(9): 1856–1860.

Ananda, J. and Herath, G. (2009). A critical review of multi-criteria decision making methods with special reference to forest management and planning. Ecological Economics, 68: 2535–2548.

Andersen, J.H., Schluter, L. and Aertebjerg, G. (2006a). Coastal eutrophication: recent developments in definitions and implications for monitoring strategies. Journal of Plankton Research, 28: 621–628.

Andersen, J.H., Axe, P., Backer, H. et al. (2010). Getting the measure of eutrophication in the Baltic Sea: towards improved assessment principles and methods. Biogeochemistry, 106: 137–156.

Andersen, S. and Wettestad, J. (1995). International problem solving effectiveness: The Oslo Project story so far. International Environmental Affairs, 7(2): 127–149.

Andersen, T., Carstensen, J. and Zardivar, J.M. (2006b). Literature data base on statistical methods for threshold identification. THRESHOLDS: Threshold of Environmental Sustainability, D2.2.1, Sixth Framework Programme; http://www.thresholds-eu.org.

Anderson, D.M. and Garrison, D.L. (1997). The ecology and oceanography of harmful algal blooms: preface. Limnology and Oceanography, 42: 1007–1009.

Ansari, A.A., Gill, S.S., Lanza, G.R. and rast, W. (Eds.). (2011). Eutrophication: Causes, Consequences and Control. Springer.

Ansari, A.A., Ghanim, S.A., Trivedi, S. et al. (2015). Seasonal dynamics in the trophic status of water, floral and faunal density along some selected coastal areas of the Red Sea, Tabuk, Saudi Arabia. International Aquatic Research, 7: 337–348.

Anttila, S., Kairesalo, T. and Pellikka, P. (2008). A feasible method to assess inaccuracy caused by patchiness in water quality monitoring. Environ. Monit. Assess, 142: 11–22.

Arfi, R., Bouvy, M. and Menard, F. (2002). Environmental variability at a Coastal Station near Abidjan: Oceanic and Continental influences. pp. 103–118. *In*: McGlade, J.M., Cury, P., Koranteng, K.A. and Hardman-Mountford, N.J. (Eds.). The Gulf of Guinea: Large Marine Ecosystem: Environmental Forcing and Sustainable Development of Marine Resources. Elsevier. Amsterdam.

Arhonditsis, G., Karydis, M. and Tsirtsis, G. (2003). Analysis of phytoplankton community structure using similarity indices: a new methodology for discriminating among eutrophication levels in coastal marine ecosystems. Environmental Management, 31: 619–632.

ASCLME/SWIOFP. (2012). Transboundary Diagnostic Analysis for the Western Indian Ocean. Volume I: Baseline. South Africa.

Baan, P.J.A. and van Buuren, J.T. (2003). Testing indicators for the marine and coastal environment in Europe. Part 3: Present state and development of indicators for eutrophication, hazardous substances, oil and ecological quality. European Environment Agency, Copenhagen. Technical Report No. 86.

Backer, H. and Leppanen, J.M. (2008). The HELCOM system of a vision, strategic goals and ecological objectives: implementing ecosystem approach to the management of human activities in the Baltic Sea. Aquatic Conservation: Marine and Freshwater Ecosystems, 18: 321–334.

Backer, L.C., McNeel, S.V., Barker, T. et al. (2010). Recreational exposure to microcystins during algal blooms in two California Lakes. Toxicon, 55: 909–921.

Baden, S.P., Loo, O., Pihl, L. and Rosenberg, R. (1990). Effects of eutrophication on benthic communities including fish: Swedish west coast. Ambio, 19: 113–122.

Bakan, G. and Buyukgongur, H. (2000). The Black Sea. Marine Pollution Bulletin, 41: 24–43.

Bald, J., Borja, A., Muxica, I., Franco, V. and Valencia, V. (2005). Assessing reference conditions and physico-chemical status according to the European Water Framework Directive: a case study from the Basque Country (Northern Spain). Marine Pollution Bulletin, 50: 1508–1522.

Balkas, T. et al. (1990). State of the marine environment in the Black Sea Region. UNEP Regional Seas Report and Studies No. 124.

Bandeira, S.O. (2011). Seagrasses: in a field guide to the seashores of Eastern Africa and the Western Indian Ocean Island (Richmond, M.D., ed.), 3rd Edition. pp. 74–77.

Barale, V., Jaquet, J.M. and Ndiaye, M. (2008). Algal blooming patterns and anomalies in the Mediterranean Sea as derived from the SeaWiFS data set (1998–2003). Remote Sensing Environment, 112: 3300–3313.

Barlow, R., Lamont, T., Kyewalyana, M., Sessions, H., van den Berg, M. and Duncan, F. (2011). Phytoplankton production and adaptation in the vicinity of Pemba and Zanzibar Islands, Tanzania. African Journal of Marine Science, 33: 283–295.

Barnett, V. and Lewis, T. (1994). Outliers in Statistical Data. J. Wiley & Sons. Chichester.

Bartleson, R.D., Kemp, W.M. and Stevenson, J.C. (2005). Use of a simulation model to examine effects of nutrient loading and grazing on *Potamogeton perfoliatus* L. communities in microcosms. Ecological Modeling, 185: 483–512.

Basti, L., Uchida, H., Matsushima, R. et al. (2015). Influence of temperature on growth and production of pectenotoxin-2 by amonoclonal culture of *Dinophysis caudata*. Marine Drugs, 13: 7124–7137.

Basurco, B. and Lovatelli, A. (2016). The aquaculture situation in the Mediterranean Sea: prediction for the future. Available at: http://www.researchgate.net/publication/266498924 Accessed 14 January 2019.

Bazzoni, A.M., Mudadu, A.G., Lorenzioni, G. et al. (2018). Detection of *Dinophysis* species and associated okadic acid in farmed shellfish: a two-year study from the western Mediterranean area. Journal of Veterinary Research 62: 137–144.

Becher, K.D., Schnoebelen, D.J. and Akers, K.K.B. (2000). Nutrients discharged to the Mississippi River from eastern Iowa watersheds, 1996–1997. Journal of the American Water Resources Association, 36: 161–173.

Becker, G.A. (1981). Contributions to the hydrography and heat budget of the North Sea. Dutch Hydrography Z34: 167–262.

Beckman, R.J. and Cook, R.D. (1983). Outliers. Technometrics, 25(2): 119–163.

Belin, C., Berthone, J.P. and Lassus, p. (1989). Dinoflagellates toxique et phenomena d' eaux colores sur les cotes francaise: evolution entre 1972 et 1988. Hydroecology, 1-2: 3–17.

Bell, S., McGillivary, D. and Pedersen, O.W. (2008). Environmental Law. 8th edition. Oxford University Press. Oxford.

Benedict, R.E. (1986). Global Environmental Change: the international perspective. *In*: EPA and UNEP, Effects of Changes in Stratospheric Ozone and Global Climate, Volume 1.

Bennekom, A.J. van, Gieskes, W.W.C. and Tijssen, S.B. (1975). Eutrophication of Dutch coastal waters. Proceedings of the Royal Society (Series B), 189: 359–437.

Bergametti, G., Remoudaki, E., Losno, R., Steiner, E., Chatenet, B. and Buat-Menard, P. (1992). Source, transport and deposition of atmospheric phosphorus over the northwestern Mediterranean. Journal of Atmospheric Chemistry, 14: 501–513.

Bernoulli, D. (1961). The most probable choice between several observations and the formation therefrom of the most likely induction. Biometrika, 48: 3–18.

Bethoux, J.P. (1980). Mean water fluxes across sections in the Mediterranean Sea evaluated on the basis of water and salt budgets and of observed salinities. Oceanologica Acta, 3: 135–142.

Bethoux, J.P., Morin, P., Madec, C. and Gentilli, B. (1992). Phosphorus and nitrogen behavior in the Mediterranean Sea. Deep Sea Research, 39: 1641–1654.

Bethoux, J.P., Gentili, B. and Tailliez, D. (1998). Warming and freshwater budget change in the Mediterranean since the 1940s, their possible relation to the greenhouse effect. Geophysical Research Letters, 25: 1023–1026.

Beyers, R.J. and Odum, H.T. (1993). Ecological Microcosms. Springer-Verlag. 557pp.

Birnie, P., Boyle, A. and Redgwell, C. (2009). International Law and the Environment. 3rd Edition. Oxford University Press. Oxford. 849pp.

Blanchet, F.G., Legendre, P. and Borcard, D. (2008). Modelling directional spatial processes in ecological data. Ecological Modelling, 215: 325–336.

Blanco, E.P., Karlsson, C., Pallon, J. et al. (2015). Cellular nutrient content measured with the nuclear microprobe and toxins produced by *Dinophysis norvegica* (Dinophyceae) from the Trondheim fjord (Norway). Aquatic Microbial Ecology, 75: 259–269.

Blink, K.H., Halpern, D., Huyer, A. and Smith, R.L. (1983). The physical environment of the Peruvian upwelling system. Progress in Oceanography, 12: 285–305.

Boehmer-Christiansen, S. (1994). The role of science in international regulation of pollution. pp. 143–167. *In*: Andersen, S. and Ostreng, W. (Eds.). International Resource Management: The Role of Science and Politics. Belhaven Press, London, New York.

Boelsch, D.F. (1977). Application of numerical classification in ecological investigations of water pollution. U.S. Environmental Research Agency, EPA-600/3-77-033. March. Science of the Total Environment. Elsevier, Sppl. 4: 419–426.

Bolam, S.G. and Fernandes, T.F. (2002). The effects of macroalgal cover on the spatial distribution of microbenthic invertebrates: the effect of macroalgal morphology. Hydrobiologia, 475: 437–448.

Boni, L., Mancini, A., Milandri, A., Poletti, R., Pompei, M. and Viviani, R. (1992). First cases of DPS in the Northern Adriatic Sea. pp. 21–24. *In*: Vollenweider, R.A., Manchetti, R. and Viviani, R. (Eds.). Marine Coastal Eutrophication. Proceedings of an International Conference, Bologna, Italy.

Bonsdorff, E., Blomquist, E.M., Mattila, J. and Norko, A. (1997). Long-term changes and coastal eutrophication. Examples from the Aland Islands and the Archipelago Sea, Northern Baltic Sea. Oceanologica Acta, 20: 319–329.

Borja, A. Franco, J. and Perez, V. (2000). A marine biotic index to establish ecological quality of soft-bottom benthos within European estuarine and coastal environments. Marine Pollution Bulletin, 40(12): 1100–1114.

Borja, A., Franco, J. and Muxica, I. (2003). Classification tools for marine ecological quality assessment: the usefulness of microbenthic communities in an area affected by a submarine outfall. ICES CM J02, 1–10.

Borja, A., Galparsono, I., Solaun, O. et al. (2006). The European Waster Framework Directive and the DPSIR, a methodological approach to assess the risk of failing to achieve good ecological status. Estuarine Coastal and Shelf Science, 66: 84–96.

Borja, A., Josefson, A.B., Miles, A. et al. (2007). An approach to intercalibration of benthic ecological status assessment in the North Atlantic ecoregion, according to the European Water Framework Directive. Marine Pollution Bulletin, 55: 42–52.

Borja, A. and Dauer, D.M. (2008). Assessing the environmental quality status in estuarine and coastal systems: comparing methodologies and indices. Ecological Indicators, 8: 331–337.

Borja, A., Dauer, D.M. and Gremare, A. (2012). The importance of setting targets and reference conditions in assessing marine ecosystem quality. Ecological Indicators, 12: 1–7.

Borlado, A.A. and Vieira, M.E.C. (2005). Spatial variability of phytoplankton, bacteria and viruses in the mesotidal salt wedge Douro Estuary (Portugal). Estuarine, Coastal and Shelf Science, 63: 143–154.

Boroumand, A., Rajaee, T. and Masoumi, F. (2018). Semivariance analysis and transinformation entropy for optimal redesigning of nutrients monitoring network in San Francisco bay. Marine Pollution Bulletin, 129(2): 689–694.

Botes, L., Sym, S.D. and Pitcher, G.C. (2003). *Karenia cristata* sp. nov. and *Karenia bicuneiformis*, sp. nov. (Gymnodiniales, Dinophyceae): two new Karenia species from the South African coast. Phycologia, 42: 563–571.

Bower, B.T. and Kneese, A.V. (1968). Managing Water Quality: Economics, Technology, Institutions. Johns Hopkins University Press, Balitmore, Maryland.

Bravo, I., Reguera, B., Martinez, A. and Fraga, S. (1990). First report of *Gymnodinium catenatum* Graham in the Mediterranean Coast. pp. 449–452. *In*: Graneli, E., Sandstorm, B., Elder, L. and Anderson, D.M. (Eds.). Toxic Marine Phytoplankton. Elsevier Science, New York.

Bresciani, M., Adamo, M., De Carolis, G., Matta, E., Pasquariello, G., Vaičiūtė, D. and Giardino, C. (2014). Monitoring blooms and surface accumulation of cyanobacteria in the Curonian Lagoon by combining MERIS and ASAR data. Remote Sensing of Environment, 146: 124–135.

Brito, A.C., Brotas, V., Caetano, M. et al. (2012). Defining phytoplankton class boundaries in Portuguese transitional waters: an evaluation of the ecological quality status according to the Water Framework Directive. Ecological Indicators, 19: 5–14.

Brito, A.C., Silva, T., Beltran, C., Chainho, P. and de Lima, R.F. (2017). Phytoplankton in two tropical mangroves of Sao Tome Island (Gulf of Guinea): a contribution towards sustainable management strategies. Regional Studies in Marine Science, 9: 89–96.

Brodie, J. (1999). Management of the Great Barrier Reef as a large marine ecosystem. *In*: Sherman, K. and Tang, O. (Eds.). Large Marine Ecosystems of the Pacific Rim: Assessment, Sustainability and Management. Blackwell Science, Mass.

BSC. (2008). State of the environment of the Black Sea (2001–2006/7). *In*: Temel Oguz (Ed.). Publications of the Commission on the Protection of the Black Sea Against Pollution (BSC). Istanbul. 448pp.

BSC. (2009a). The Convention on the Protection of the Black Sea Against Pollution. Retrieved from: http://www.blacksea-commission.org/_convention.asp.

BSC. (2009b). Implementation of the Strategic Action Plan for the Rehabilitation and Protection of the Black Sea Against Pollution (BSC), 2008–2013, Istanbul, Turkey. Available at: http://www.blacksea.commission.org.

Burger, J. (2006). Bioindicators: a review of their use in the environmental literature 1970–2005. Environmental Bioindicators, 1(2): 136–144.

Burkepile, D.E. and Hay, M.E. (2006). Herbivore vs nutrient control of marine primary producers: context dependent effects. Ecology, 87: 3128–3139.

Burkholder, J.M., Noga, E.J., Hobbs, C.H. and Glasgow, H.B. (1992). New "phantom" dinoflagellate is the causative agent of major estuarine fish kills. Nature, 358: 407–410.

Burrough, P.A. and McDonnell, R.A. (2000). Principles of Geographic Information Systems: Spatial Information Systems and Geostatistics. Oxford: Oxford University Press.

Butler, M.J., Hunt, J.H. and Herrnkind, W.F. (2005). Cascading disturbances in Florida Bay, USA: cyanobacteria blooms, sponge mortality and implications for juvenile spiny lobsters Panulirus argus. Marine Ecology Progress Series, 129: 119–125.

Buyakates, Y. and ·Roelke, D. (2005). Influence of pulsed inflows and nutrient loading on zooplankton and phytoplankton community structure and biomass in microcosm experiments using estuarine assemblages. Hydrobiologia, 548: 233–249.

Caldwell, L.K. (1996). International Environmental Policy. From the Twentieth to Twenty-First Century. 3rd Edition. Duke University Press, Durham and London. 484pp.

Cao, Y., Bark, A.W. and Williams, W.P. (1997). A comparison of clustering methods for river benthic community analysis. Hydrobiologia, 347: 25–40.

Carbiener, R. (1992). Compositions lessivielles avec eau sans phosphates et protection des milieux aquatiques. Rapport au secretaire d'etat aupres du premier minister, charge de l' environment. 182pp.

Carlson, R.E. (1977). A trophic state index for lakes. Limnology and Oceanography, 22(2): 361–369.

Carrada, G.C., Casotti, R. and Saggiorno, V. (1988). Occurrence of a bloom of Gymnodinium catenatum in a Tyrrhenian coastal lagoon. Rapports de la Commission International de la Mer Mediterranee, 31(2): 61.

Carstensen, J., Heriksen, O.D. and Teilmann, J. (2006). Impacts of offshore wind farm construction on harbor porpoises: acoustic monitoring of echo-location activity using porpoise detectors (T-PODs). Marine Ecology Progress Series, 321: 295–308.

Castilla, J.C. and Nealler, E. (1978). Marine Environmental Impact due to mining activities of El Salvador Copper Mine, Chile. Marine Pollution Bulletin, 9(3): 67–70.

Ceccarelli, R. and Di Bitetto, M. (1996). La valutazione di impatto ambientale in sistemi di acquacoltura e techniche di controllo dei nutrienti. Biologia Marina Mediterranea, 3: 398–399.

CEDI. (2000). Comprehensive conservation and management plan (CCMP) for the San Juan Bay Estuary, Caribbean Environment and Development Institute, San Juan, Puerto Rico.

Celliers, L. and Ntombela, C. (2015). Urbanization, coastal development and vulnerability and catchments. pp. 387–408. *In*: UNEP-Nairobi Convention and WIOSMA: The Regional State of the Coast Report: Western Indian Ocean. UNEP and WIOSMA, Nairobi, Kenya, 546pp.

CEP. (2010). Marine and Coastal Issues. Caribbean Environment Programme. Retrieved from: http://www.cep.unep.org/publications-and-resources/marine-and-coastal-issues-links.

Chao, S., Kao, T. and Al-Hajri, K. (1992). A numerical investigation of circulation in the Arabian Gulf. Journal of Geophysical Research, 97: 219–236.

Chapman, A.R.O. and Craigie, J.S. (1977). Seasonal growth in Laminaria longicirrus: relations with dissolved inorganic nutrients and internal reserves of nitrogen. Marine Biology, 40: 197–205.

Chiaudani, G., Marchetti, R. and Vighi, M. (1980). Eutrophication in Emilia-Romagna coastal waters (North Adriatic Sea, Italy): a case history. Progress in Water Technology, 12pp.

Chislock, M.F., Doster, E., Zitoner, R.A. and Wilson, A.E. (2013). Eutrophication: causes, consequences and controls in aquatic ecosystems. Nature Education Knowledge.

Cho, B.C. and Azam, F. (1990). Biochemical significance of bacterial biomass in the ocean's euphotic zone. Marine Ecology Progress Series, 63: 253–259.

Chou, L.M. (1994). Marine environmental issues of Southeast Asia: state and development. Hydrobiologia, 285: 139–150.

Christodoulaki, S., Petihakis, G., Kanakidou, M. et al. (2013). Atmospheric deposition in the Eastern Mediterranean. A driving force for ecosystem dynamics. Journal of Marine Systems, 109-110: 78–93.

Clark, R.B. (1971). Futility of phosphate detergent Ban. Marine Pollution Bulletin, 2(4): 50–51.

Clark, R.B. (2001). Marine Pollution, 3rd Edition, Oxford University Press, Oxford.

Clarke, K.R. (1990). Comparisons of dominance curves. Journal of Experimental Marine Biology and Ecology, 138: 143–157.

Cliff, A.D. and Ord, J.K. (1973). Spatial Autocorrelation (Monographs in Spatial and Environmental Systems Analysis). London: Pion Ltd.

Cloern, J.E. (2001). Our evolving conceptual model of the coastal eutrophication problem. Marine Ecology-Progress Series, 210: 223–253.

Coffin, M.R.S., Courtenay, S.C., Pater, C.C. and van den Heuvel, M.R. (2018). An empirical model using dissolved oxygen as an indicator for eutrophication at a regional scale. Marine Pollution Bulletin, 133: 261–270.

Colella, S., Falcini, F., Rinaldi, E., Sammartino, M. and Santoleri, R. (2016). Mediterranean ocean colour chlorophyll trends. PLoS ONE 11(6): e0155756. https://doi.org/10.1371/journal.pone.0155756.

Collingridge, D. and Reeve, C. (1986). Science Speaks To Power. The Role Of Experts In Policy Making. Pinter, London.

Colmenares, N.A. and Escobar, J.J. (2002). Ocean and Coastal issues and policy responses in the Caribbean. Ocean and Coastal Management, 45: 905–924.

Corredor, J.E. and Morell, J.M. (2001). Seasonal variations of physical and biochemical features in eastern Caribbean surface waters. Journal of Geophysical Research, 106. N C3: 4517–1525.

Cosme, N., Koski, M. and Hauschild, M.Z. (2015). Exposure factors for marine eutrophication impacts assessment based on a mechanistic biological model. Ecological Modelling, 317: 50–63.

Costanza, R., d'Arge, R., de Grod, R. et al. (1997). The value of the world's ecosystem services and natural capital. Nature, 387: 253–260.

Costanza, R., de Groot, R., Sutton, P. et al. (2014). Changes in the global value of ecosystem services. Global Environmental Change, 26: 152–158.

Couper, A. (1983). Times Atlas of the Oceans. Times Books, Ltd., London.

Cressie, N.A.C. (1993). Statistics for Spatial Data. New York: Wiley.

Crossland, C.J., Kremer, H.H., Lindeboom, H.J., Crossland, J.I.M. and Le Tissier, M.D.A. (2005). Coastal Fluxes in the Anthropocene. Berlin: Springer.

Cruzado, A. (1985). Chemistry of Mediterranean waters. pp. 126–147. *In*: Margalef, R. (Ed.). Western Mediterranean, Pergamon Press, Oxford.

Cruzet, P., Nixon, S., Rees, Y., Parr, W., Laffon, L., Bogestrand, J., Kristensen, P., Lallana, C., Izzo, G., Bokn, T., Bak, J., Lack, T.J. and Thyssen, N. (1999). Nutrients in European ecosystems. Environmental Assessment Report, 4, EEA, Copenhagen, 155pp. Available from http://www.eea.eu.int.

DAFFE/IME. (2000). Annex. The Guidelines propose that enterprises should, in the countries in which they operate, contribute to 'economic, social and environmental progress with the view to achieving sustainable development' (General, Paragraph 1).

Daily, G. (1997). Nature's Services: Societal Dependence on Natural Ecosystems. Island Press, Washington.

Danilov, R.A. and Ekelund, N.G.A. (2001). Comparative studies on the usefulness of seven ecological indices for the marine coastal monitoring close to the shore on the Swedish East coast. Environmental Monitoring and Assessment, 66: 265–279.

Danovaro, R. (2003). Pollution threats in the Mediterranean Sea: an overview. Chemistry and Ecology, 19: 15–32.

daNUbs. (2005). Nutrient Management in the Danube Basin and its impact on the Black Sea. Final Report pp69. Available at: http://danubs.tuwien.ac.at.

Day, R.W. and Quinn, G.P. (1989). Comparisons of treatments after an analysis of variance in ecology. Ecological Monographs, 59: 433–463.

de Jong, F. (2006). Marine eutrophication in Perspective: on the Relevance of Ecology for Environmental Policy. Springer. Berlin. 335pp.

de Jong, V.N., Elliot, M. and Orive, E. (2002). Causes, historical development, effects and future challenges of a common environmental problem: eutrophication. Hydrobiologia, 475: 1–19.

de Leo, G.A., Bartoli, M., Naldi, M. and Viaroli, P. (2002). A first generation stochastic bioeconomic analysis of algal bloom control in a coastal lagoon (Sacca id Goro, Po River Delta). Marine Ecology, 23: 92–100.

Delgado, M., Estrada, M., Camp, J., Fernadez, J.V., Santamarti, M. and Lleti, C. (1990). Development of a toxic *Alexandrium minutum* Halim (Dinophyceae) bloom in the harbor of Sant Carles de la Rapita (Ebro Delta Northwestern Mediterranean). Scientia Marina, 54: 1–7.

Deo Florence, L.O., Arturo, O.L. and Rhodora, V.A. (2014). Development, morphological characteristics and viability of temporary cysts of *Pyrodinium Bahamense* var. *compressum* (Dinophyceae) *in vitro*. European Journal of Phycology, 49(3): 265–275.

Desai, B.H. (2006). UNEP: a global environmental authority. Environmental Policy and Law, 36: 3–4.

Desmit, X., Thieu, V., Billen, G., Campuzano, F., Duliere, V., Garnier, J., Lassaletta, L., Menesguen, A., Neves, R., Pinto, L., Silvestre, M., Sobrinho, J.L. and Lacroix, G. (2018). Reducing marine eutrophication may require a paradigmatic change. Science of the Total Environment, 635: 1444–1466.

Diaz, J.M. and Acero, A. (2003). Marine biodiversity in Colombia: achievements, status of knowledge and challenges. Gayana, 67(2): 261–274.

Diaz, R.J. and Rosenberg, R. (2008). Spreading dead zones and consequences for marine ecosystems. Science, 321: 926–929.

Dicks, B., Bakke, T. and Dixon, I.M.T. (1988). Oil exploration and production of oil spills. pp. 524–537. *In*: Salomons, W., Bayne, B.L., Duursma, E.K. and Forstrner, U. (Eds.). Pollution of the North Sea: An Assessment. Springer-Verlag.

Digby, P.G.N. and Kempton, R.A. (1987). Multivariate Analysis of Ecological Communities. Chapman and Hall. London. 206pp.

DiMento, J. (1995). Hazardous waste management in the Middle East: some confidence buildings in the struggle for the peace. *In*: Spiegel, S. (Ed.). Practical Peacemaking in the Middle East: The Environment, Water, Refugees and Economic Cooperation and Development. Garland, New York.

DiMento, J.F.C. (2003). The Global Environment and International Law. University of Texas Press, Austin.

DiMento, J.F.C. and Hickman, A.J. (2012). Environmental Governance of the Great Seas: Law and Effect. Edward Elgar, Cheltenham, UK. 220pp.

Doering, P.H., Oviatt, C.A., Beatty, L.L., Banzon, V.F., Rice, R., Kelly, S.P., Sullivan, B.K. and Frithsen, J.B. (1990). Structure and function in a model coastal ecosystem; silicon, the benthòs and eutrophication. Marine Ecology Progress Series, 52: 287–299.

Dolberth, M., Pardal, M.A., Lillebo, A.I. et al. (2003). Short and long term effects of eutrophication, the secondary production of an intertidal microbenthic community. Marine Biology, 143: 1229–1238.

Dorantes-Aranda, J.J., Seger, A., Mardones, J.I. et al. (2015). Progress in understanding algal bloom-mediated fish kills: the role of superoxide radicals, phycotoxins and fatty acids. PLOS ONE 10(7): e0133549.

Douglass, M. (2000). Mega-urban regions and World City formation: globalization and economic crisis and urban policy issues in Pacific Asia. Urban Studies, 37: 2315–2335.

Douvere, F. and Ehler, C.N. (2011). The importance of monitoring and evaluation in adaptive marine spatial planning. Journal of Coastal Conservation, 15: 305–311.

Draper, N.R. and Smith, H. (1998). Applied regression analysis. Wiley Series in Probability and Statistics. J. Wiley and Sons Inc.

Duarte, C.M. (2009). Coastal eutrophication research: a new awareness. Hydrobiologia, 629: 263–269.

Duce, R.A., Liss, P.S., Merrill, J.T. et al. (1991). The atmospheric input of trace species to the world ocean. Global Biogeochemical Cycles, 5: 193–259.

Duce, R.A., LaRoch, J., Altieri, K. et al. (2008). Impacts of atmospheric nitrogen on the open ocean. Science, 320: 893–897.

Duce, R.A. (2009). The impacts of atmospheric deposition to the oceans on marine ecosystems and climate.

Ducrotoy, J.P. (1999). Indications of change in the marine flora of the North Sea in the 1990s. Marine Pollution Bulletin, 38: 646–654.

Dugdale, R.C. (1967). Nutrient limitation in the sea: dynamics, identification and significance. Limnology and Oceanography, 12: 685–695.

Dunteman, G.H. (1989). Principal Component Analysis. Sage Publications. London.

Dutilleul, P. (1993). Spatial heterogeneity and the design of ecological field experiments. Ecology, 74: 1646–1658.

EC. (1991a). Council Directive 91/271/EEC of 21 May 1991 concerning urban waste-water treatment. OJ L135, 30.5.

EC. (1991b). Council Directive concerning the protection of waters against pollution caused by nitrates from agricultural sources (91/676/EEC). Off. J. Eur. Commun., L375(1-8): 31.12.

EC. (2000). European Commission Directive 2000/60/EC of the European Parliament and of the Council of 23 October 2000 establishing a framework for community action in the field of water policy. Brussels Off. Eur. Commun. L327.

EC. (2006). Directive 2006/7/EC of the European Parliament and the Council of 15 February 2006 concerning the management of bathing water quality and repealing Directive 76/160/EEC. Official Journal of the European Community, L64: 37–51.

EC. (2008). Directive 2008/56/EC of the European Parliament and of the Council of 17 June 2008 on establishing a framework for community action in the field of marine environment policy (Marine Strategy Framework Directive). L164/19.

ECOSTAT. (2004). Conceptual framework for eutrophication assessment in the context of European Water Policies. Report for Eutrophication Steering Group.

EEA. (1999). Nutrients in European Ecosystems. Environmental assessment report No. 4. European Environment Agency, Copenhagen.

EEA. (2001). Eutrophication in Europe's coastal waters. Topic Report No. 7. Copenhagen: European Environmental Agency.

EEC. (1979). Council Directive of 30 October 1979 on the quality required of shellfish waters (79/923/EEC). Official Journal of the European Community, L281: 48.

EEC. (1992). Council Directive 92/43/EEC of 21 may 1992 on the conservation of natural habitats and of wild fauna and flora. Official Journal of the European Community. L43: 1–66.

Eker, E., Georgieva, L., Senichlina, L. and Kideys, A.E. (1999). Phytoplankton distribution in the western and eastern Black Sea in spring and autumn 1995. ICES Journal of Marine Science, 56 Supplement: 15–22.

Elbersen, W., Wiersinga, R. and Waarts, Y. (2005). Market scan bioenergy Ukraine. Agrotechnology and food innovations in B.V. Wageningen: Report BO-10-006-062; 200952. Available at: http://www.biomassandbioenergy.nl/bioenergyukraine.htm.

Eleveld, M.A. and van der Woerd, H.J. (2006). Patterns in water quality products of the North Sea: variogram analyses of single and compound SeaWiFS CHL & SPM grids. In: Kerle, N. and Skidmore, A.K. (Eds.). Remote Sensing: From Pixels to Processes. The Netherlands: Ensche, ISPRS Proceedings Technical Commission, Vol. VII.

Elliot, J.M. (1983). Some methods for the statistical analysis of samples from Benthic Invertebrates. Scientific Publication No. 25. Freshwater Biological Association. Ferry House, Ambleside, Cumbria, UK.

Elliot, M. (1994). The analysis of microbenthic community data. Marine Pollution Bulletin, 28(2): 62–64.

Engelhardt, H.T. and Caplan, A.L. (1987). Patterns of controversy and closure: the interplay of knowledge, values and political forces. pp. 1–23. In: Engelhardt, H.T. and Caplan, A.L. (Eds.).

Scientific Controversies. Case Studies in the Resolution and Closure of Disputes in Science and Technology. Cambridge University Press, Cambridge.

EPA. (1998). Tier 2 Report to Congress US. Government Printing Office, Washington D.C.

ESCAP. (2009). Economic and Social Commission for Asia and the Pacific: Statistical Yearbook for Asia and the Pacific, United Nations Publications.

Estrada, M., Marrace, C., Latasa, M., Berdalet, E., Delgado, M. and Riera, T. (1993). Variability of deep chlorophyll maximum characteristics in the Northwestern Mediterranean. Marine Ecology Progress Series, 92: 289–300.

Estrada, M., Berdalet, E., Marrase, C., Arin, L. and McLean, M. (1996). Effect of different nutrient combinations on phytoplankton development in microcosms. pp. 297–300. *In*: Yasumoto, T., Oshima, Y. and Fukuyo, Y. (Eds.). Harmful and Toxic Algal Blooms. Intergovernmental Oceanographic Commission of UNESCO. Paris.

Estrada, M. and Peters, F. (2002). Microcosms: applications in marine phytoplankton studies. pp. 359–370. *In*: Subba Rao, D.V. (Ed.). Pelagic Ecology Methodology. A.A. Balkema Publishers. Lisse. 464pp.

Everard, M. (2017). Ecosystem Services: Key Issues. Routledge Taylor and Francis Group, London. 188pp.

Everitt, B.S. (1981). A Monte-Carlo investigation of the likelihood ratio test for the number of components in a mixture of normal distributions. Multivariate Behavioral Research, 16(2): 171–180.

Everitt, B.S., Landau, S. and Leese, M. (2001). Cluster Analysis. Arnold. London. 237pp.

Fabricus, K.E. (2005). Effects of terrestrial run off on the ecology of corals and coral reefs: review and synthesis. Marine Pollution Bulletin, 50: 125–146.

Fanning, A.K. (1989). Influence of atmospheric pollution on nutrient limitation in the ocean. Nature, 339: 460–463.

Fanning, L. et al. (2007). A large marine ecosystem framework. Marine Policy, 32: 434–443.

FAO. (2005). Fertilizer use by Crop in Ukraine: Land and Plant Nutrition Management Service, Land and Water Development Division. Rome, Italy Food and Agriculture Organization of the United Nations.

Fehling, J., Green, D.H. and Davidson, K. (2004). Domoic acid production by *Pseudo-nitzschia seriata* (Bacilariophyceae) in Scottish waters. Journal of Phycology, 40: 622–630.

Feki, W., Hamza, A., Frossard, V. et al. (2013). What are the potential drivers of blooms of the toxic dinoflagellate *Karenia selliformis*? A 10-year study in the Gulf of Gabes, Tunisia, southwestern Mediterranean Sea. Harmful Algae, 23: 8–18.

Fennery, S. and Green, A. (2015). Shelf sediments and biodiversity. *In*: UNEP-Nairobi Convention and WIOSMA: The Regional State of the Coast Report: Western Indian Ocean. UNEP and WIOSMA, Nairobi, Kenya, 546pp.

Ferreira, J.G., Andersen, J.H., Borja, A. et al. (2011). Overview of eutrophication indicators to assess environmental status within the European Marine Strategy Framework Directive. Estuarine, Coastal and Shelf Science, 93: 117–131.

Ferris, R. and Humphrey, J.W. (1999). A review of potential biodiversity indicators for application in British Forests. 72(4): 313–328.

Fitzmaurice, M. (1998). The Helsinki Conventions 1974 and 1992. IJMCL, 13: 379–399.

Fleming-Lehtinen, V. (2007). HELCOM EUTRO: Development of tools for a Thematic Eutrophication Assessment for two Baltic Sea sub-regions, the Gulf of Finland and Bothnian Bay. Volume 13.

Flindt, M.R., Pardal, M.A., Lillebo, A.I. et al. (1999). Nutrient cycling and plant dynamics in estuaries: a brief overview. Acta Oceanologica, 4: 237–248.

Fogg, G.E. and Thake, B. (1987). Algal Cultures and Phytoplankton Ecology. 3rd Edition. The University of Wisconsin Press. 269pp.

Foster, J. and Hart, S. (Eds.). (2016). Global Aquaculture Alliance. Harmful Algal Blooms. Assessing Chile's Historic HAB Events of 2016. Available at: http://www.aquaculturealliance.org, retrieved on June 6th, 2019.

Fowler, J., Cohen, L. and Jarvis, P. (1998). Practical Statistics for Field Biology. J. Wiley & Sons, Chichester.

Francis, J. and Torell, E. (2004). Human dimensions of coastal management in the Western Indian Ocean region. Ocean and Coastal Management, 47: 299–307.

Frank, V. (2007). The European Community and Marine Environmental Law of the Sea: Implementing Global Obligations at the Regional Level. Martianus Nijhoff Publishers. Leiden.

Gallisai, R., Peters, F., Volpe, G. et al. (2014). Saharan dust deposition may affect phytoplankton growth in the Mediterranean sea at ecological time scales. PLOS ONE, 9(10): e110762.

Garcia, C.A.E., Garcia, V.M.T. and McClain, C.R. (2005). Evaluation of SeaWiFS chlorophyll algorithms in the Southwestern Atlantic and Southern Oceans. Remote Sensing of Environment, 95: 125–137.

Garmendia, M., Borja, A., Breton, F. et al. (2015). Challenges and difficulties in assessing environmental status under the requirements of the Ecosystem Approach in North African countries, illustrated by eutrophication assessment. Environmental Monitoring and Assessment, 187: 289.

Garmendia, M., Borja, A., Franco, J. and Revilla, M. (2013). Phytoplankton composition indicators for the assessment of eutrophication in marine waters: present state and challenges within the European directives. Marine Pollution Bulletin, 66: 7–16.

Garpe, K. (2008). Ecosystem services provided by the Baltic Sea and Skagerrak. Swedish Environmental Protection Agency. Stockholm, Sweden, pp193.

Garzón-Ferreira, J., Cortes, J., Croquer, A., Guzmán, H., Leao, Z. and Rodríguez-Ramirez, A. (2000). Status of coral reefs in southern tropical America: Brazil, Colombia, Costa Rica, Panamá and Venezuela. In: Wilkinson, C. (Ed.). Status of Coral Reefs of the World: Australian Institute of Marine Sciences (AIMS) 2000, Australia.

Gattuso, J.P., Gentili, B., Duarte, C.M. et al. (2006). Light availability in the coastal ocean: impact of the distribution of benthic photosynthetic organisms and their contribution to primary production. Biogeosciences, 3: 489–513.

Gauch, H.G. (1989). Multivariate Analysis in Community Ecology. Cambridge University Press. Cambridge. 298pp.

Georgopoulos, P.G. and Seinfeld, J.H. (1982). Statistical distributions of air pollutant concentrations. Environmental Science and Technology, 16: A401–A416.

Gerati, J.R., Anderson, D.M., Timperi, R.J. et al. (1989). Humpback whales (Megaptera novaeangliae) fatally poisoned by dinoflagellate toxin. Canadian Journal of Fisheries and Aquatic Sciences, 46: 1895–1898.

GESAMP. (1990). The state of the marine environment. United Nations Environment Programme. UNEP Regional Seas Reports and Studies, 115: 111.

Giancarlo, R. and Utro, F. (2012). Algorithmic paradigms for stability based cluster validity and model selection statistical methods, with applications to microarray data analysis. Theoretical Computer Science, 428: 58–79.

Giordani, G., Zardivar, J.M. and Vialori, P. (2009). Simple tools for assessing water quality and trophic status in transitional water ecosystems. Ecological Indicators, 9(5): 982–991.

Giovanardi, F. and Tromellini, E. (1992). Statistical assessment of trophic conditions: application of the OECD methodology to the marine environment. pp. 211–233. In: Vollenweider, R.A., Marchetti, R. and Viviani, R. (Eds.). Marine Coastal Eutrophication. Elsevier Science Publications, London. 1310pp.

Giupponi, C., Eiselt, B. and Ghetti, P.F. (1999). A multicriteria approach for mapping risks of agricultural pollution for water resources: The Venice Lagoon Watershed case study. Journal of Environmental Management, 56: 259–269.

Gladstone, W. (2000). Ecological and social basis for management of the Red Sea marine protected areas. Ocean and Coastal Management, 43: 1015–1032.

Glen, I.M., Soderqvist, T. and Wulff, F. (1997). Nutrient reductions to the Baltic Sea. Ecology, costs and benefits. Journal of Environmental Management, 51: 123–143.

Gochfeld, M. (1980). Mercury levels in some seabeds of the humbolt current, Peru. Environmental Pollution, A22: 197–205.

Gohin, F., Saulquin, B., Oger-Jeanneret, H., Lozach, L., Lampert, L., Lefebvre, A., Riou, P. and Bruchon, F. (2008). Towards a better assessment of the ecological status of coastal waters using satellite-derived chlorophyll-a concentrations. Remote Sensing of Environment, 112: 3329–3340.

Goldman, J.C. and Carpenter, E.J. (1974). A kinetic approach to the effect of temperature on algal growth. Limnology and Oceanography, 19: 756–766.

Goldman, J.C., McCarthy, J.J. and Peavey, D.G. (1979). Growth rate influence on the chemical composition of phytoplankton in oceanic waters. Nature, 279: 210–215.

Goodall, D.W. (1954). Objective methods for the classification of vegetation. III. An essay in the use of factor analysis. Australian Journal of Botany, 2: 304–324.

Gossling, S. (2006). Towards sustainable tourism in the Western Indian Ocean. Western Indian Ocean Journal of Marine Science, 5: 55–70.

Govindan, K. and Jepsen, M.B. (2016). ELECTRE: A comprehensive literature review on methodologies and applications. European Journal of Operational Research, 250(1): 1–29.

Gower, J.C. and Legendre, P. (1986). Metric and Euclidean properties of dissimilarity coefficients. Journal of Classification, 3: 5–48.

GPA. (1995). Global Programme of Action for the protection of the Marine Environment from Land-Based Activities. UNEP. Washington, D.C.

Graneli, E., Wallstrom, K., Larsson, U., Graneli, W. and Elmgren, R. (1990). Nutrient limitation of primary production in the Baltic Sea area. Ambio, 19: 142–151.

Grasshoff, K. (1976). Review on hydrographical and productivity conditions in the Gulf Region. UNESCO Technical Papers in Marine Sciences, 26: 39–62.

Gray, J. (1992). Eutrophication in the sea. pp. 3–15. *In*: Colombo, G., Ferrari, I., Cecherelli, V.U. and Rossi, R. (Eds.). Marine Eutrophication and Population Dynamics. Olsen and Olsen, Denmark.

Gray, J.S. (1982). Pollution effects on marine ecosystems. Netherland Journal of Sea research, 16: 424–443.

Gray, J.S., Clarke, K.R., Warwick, R.M. and Hobbs, G. (1990). Detection of initial effects of pollution on marine benthos: an example from the Ekofish and Elfisk oil fields North Sea. Marine Ecology Progress Series, 66: 285–299.

Griffiths, C. (2005). Coral marine biodiversity in East Africa. Indian Journal of Marine Science, 34: 35–41.

Grubbs, F.E. (1969). Proceedings for detecting outlying observations in samples. Technometrics, 11: 1–21.

Guieu, C., Aumont, A., Paytan, L. et al. (2014). The significance of the episodic nature atmospheric deposition to low nutrient low chlorophyll regions. Global Biochemical Cycles. 28: 1179–1198.

Gustafsson, B.G., Sohenk, F., Blenckner, T. et al. (2012). Reconstructing the development of Baltic Sea eutrophication 1850–2006. AMBIO, 4(16): 534–548.

Haas, P.M. (1990). Saving the Mediterranean: the Politics of International Environmental Cooperation. Columbia University Press. NY. 303pp.

Haas, P.M., Keohane, R.O. and Levy, M.A. (Eds.). (1993). Institutions for the Earth: Sources of Effective International Environmental Protection. Global Environmental Accord: Strategies for Sustainability and Institutional Innovation (2nd edition), MIT Press.

Hajdu, L.J. (1981). Graphical comparison of resemblance measures in phytosociology. Vegetatio, 48: 47–59.

Hajkowicz, S. (2007). A comparison of multiple criteria analysis and unaided approaches to environmental decision making. Environmental Science & Policy, 10: 177–184.

Hajkowicz, S. and Higgins, A. (2008). Comparison of multiple criteria analysis techniques for water resource management. European Journal of Operational Research, 184: 255–265.

Hakanson, L. (2006). Suspended Particulate Matter in Lakes, Rivers and Marine Systems. The Blackburn Press, New Jersey.

Hakanson, L. and Bryhn, A.C. (2008). Eutrophication in the Baltic Sea: Present Situation, Nutrient Transport Processes. Remedial Strategies. Springer. Berlin. 261pp.

Hall, E.R., Muller, E.M., Goulet, T. et al. (2012). Eutrophication may compromise the resilience of the Red Sea coral *Stylophora pistillata* to global change. Marine Pollution Bulletin, 131: 701–711.

Hallegraeff, G.M. (2003). Harmful algal blooms: a global overview. pp. 25–49. *In*: Hallegraeff, G.M., Anderson, D.M. and Cembella, A.D. (Eds.). Manual of Harmful Marine Algae. UNESCO Publishing, 794pp.

Hallegraeff, G.M., Anderson, D.M. and Cembella, A.D. (Eds.). (2003). Manual of Harmful Marine Algae. UNESCO Publishing, 794pp.

Halser, A.D. (1947). Eutrophication of lakes by domestic drainage. Ecology, 28(4): 383–395.

Hansson, S. and Rudstam, L.G. (1990). Eutrophication and Baltic fish communities. Ambio, 19: 123–125.

Hare, S.R. and Mantua, N.J. (2000). Empirical evidence for North Pacific regime shifts in 1977 and 1989. Progress in Oceanography, 47: 103–145.

Hartog, J.A., Hinloopen, E. and Nijkamp, P. (1989). A sensitivity analysis of multicriteria choice methods—An application on the basis of the optimal site selection for a nuclear-power plant. Energy Economics, 11: 293–300.

Harvey, E.T., Kratzer, S. and Philipson, P. (2015). Satellite-based water quality monitoring for improved spatial and temporal retrieval of chlorophyll-a in coastal waters. Remote Sensing of Environment, 158: 417–430.

Hawkins, D.M. (1980). Identification of Outliers. Chapman and Hall, London.

Heileman, L.I. and Siung-Chang, A.M. (1990). An analysis of fish kills in coastal and inland waters of Trinidad and Tobago, West Indies, 1976–1990. Caribbean Marine Studies, 1: 126–136.

Heileman, S., Lutjeharms, J.R.E. and Scott, L.E.P. (2008). The Agulhas current large marine ecosystems. pp. 2–18. *In*: Sherman, K. and Hempel, C. (Eds.). UNEP Large Marine Ecosystem Report: a Perspective on Changing Conditions in LMEs of the World's Regional Seas, UNEP, United States.

Heileman, S. (2009). I-2 Guinea Current: LME #28. Large Marine Ecosystems of the World. NOOA. Accessed: http://www.lme.noaa.gov/.

Heip, C. (1995). Eutrophication and zoobenthos dynamics. Ophelia, 41: 113–136.

HELCOM. (1996). Third Periodic Assessment of the State of the Marine Environment of the Baltic Sea area 1989–1993. Executive Summary. Baltic Sea Environment Proceedings 64A.

HELCOM. (2006). Development of tools for the assessment of eutrophication in the Baltic Sea. Environ. Proceedings, No. 104, 64pp.

HELCOM and NEFCO. (2007). Economic analysis of the BSAP with focus on eutrophication. COWI, 112pp.

HELCOM. (2009). Eutrophication in the Baltic Sea. An integrated thematic assessment of the effects of nutrient enrichment in the Baltic Sea region. Baltic Sea Environment Proceedings, No. 115B. Helsinki Commission, Finland.

HELCOM. (2010). Ecosystem health in the Baltic Sea. 2003–2007. HELCOM Initial Holistic Assessment Baltic Sea Environment Proceedings, I22.

HELCOM. (2013). Review of the fifth Baltic Sea pollution load compilation for the 2013 HELCOM Ministerial Meeting. Baltic Sea Environmental Proceedings 141, 49pp.

HELCOM. (2014). Eutrophication status of the Baltic Sea 2007–2011, A concise thematic assessment. Baltic Sea Environment Proceedings, 14.

HELCOM. (2017a). First version of the "State of the Baltic Sea". Report – June 2017. Baltic Marine Protection Commission. 202pp.

HELCOM. (2017b). The integrated assessment of eutrophication. Supplementary report to the first version of the "State of the Baltic Sea Report".

Hellegraeff, G.M. and Fraga, S. (1998). Bloom dynamics of the toxic dinoflagellate *Gymnodinium catenatum*, with emphasis on Tasmanian and Spanish coastal waters. pp. 59–80. *In*: Anderson, D.M. et al. (Eds.). The Physiological Ecology of Harmful Algal Blooms. Heidelberg. Springer-Verlag (NATO Advanced Study Institute Series G. Ecological Sciences, 41.).

Herrera-Silveira, J.A. and Morales-Ojeda, S.M. (2009). Evaluation of the health status of a coastal ecosystem in southeast Mexico: assessment of water quality, phytoplankton and submerged aquatic vegetation. Marine Pollution Bulletin, 59: 72–86.

Heyman, U., Ryding, S.O. and Forsberg, C. (1984). Frequency distributions of water quality variables: relationships between mean and maximum values. Water Research, 18: 787–794.

Hill, H.W. (1973). Currents and water masses. pp. 17–42. *In*: Godberg, E.D. (Ed.). North Sea Science. MIT, Cambridge.

Hinojosa, I.A. and Thiel, M. (2009). Floating marine debris in fjords, gulfs and channels of Southern Chile. Marine Pollution Bulletin, 58: 341–350.

Hoepner, T. and Lattemann, S. (2000). Chemical impacts from seawater desalination plants—a case study of the northern Red Sea. Desalination, 152: 133–140.

Hood, R.R. and Coles, V.J. (2004). Modelling the distribution of Trichodesmium and nitrogen fixation in the Atlantic Ocean. Journal of Geophysical Research, 109: C06006.

Howard, A.F. (1991). A critical look at multiple criteria decision making techniques with reference to forestry applications. Canadian Journal of Forestry Research, 21: 1649–1659.

Hu, J., Zhang, G., Li, K., Peng, P. and Chivas, A.R. (2008). Increased eutrophication offshore Hong Kong, China during the past 75 years. Evidence from high-resolution sedimentary records. Marine Chemistry, 110: 7–17.

Hughes, C., Kamish, W., Ally, H. et al. (2012). Investigation of Effects of Projected Climate Change on Eutrophication and Related Water Quality and Secondary Impacts on the Aquatic Ecosystem. WRC Report No. 2028/1/14.

Hunte, W. and Wittenberge, M. (1992). Effects of eutrophication and sedimentation in juvenile corals II settlement. Marine Biology, 114: 625–631.

Hunter, P.R. (1998). Cyanobacterial toxins and human health. Journal of Applied Microbiology, 84: 35–40.

Hutchinson, G.E. (1967). A Treatise on Limnology. Volume II: Introduction to Lake Biology and the Limnoplankton. J. Wiley & Sons, Inc. NY. 1115pp.

Ignatiades, L., Vassiliou, A. and Karydis, M. (1985). A comparison of phytoplankton biomass parameters and their intercorrelation in Saronicos Gulf (Greece). Hydrobiologia, 128: 201–206.

Ignatiades, L., Pagou, P. and Vassiliou, A. (1986). A long term response of 6 diatom species to eutrophication. Oceanologica Acta, 9: 449–456.

Ignatiades, L., Karydis, M. and Pagou, P. (1987). Patterns of dark $^{14}CO_2$ incorporation by natural marine-phytoplankton communities. Microbial Ecology, 13(3): 249–259.

Ignatiades, L. (1990). Photosynthetic capacity at the surface microlayer during the mixing period. Journal of Plankton Research, 12: 851–860.

Ignatiades, L. and Karydis, M. (1990). Detection of phytoplankton seasonality trends based on K-dominance curves. Rapp. Comm. Int. Mer Medit. 32: 1.

Ignatiades, L., Karydis, M. and Vounatsou, P. (1992). A possible method for evaluating oligotrophy and eutrophication based on nutrient concentration scales. Marine Pollution Bulletin, 24: 238–243.

Ignatiades, L., Georgopoulos, D. and Karydis, M. (1995). Description of the phytoplanktonic community of the oligotrophic waters of the SE Aegean Sea (Mediterranean). Mar. Ecol. Pubbl. Della Stn Zoo Di Napoli I, 16: 13–26.

Ignatiades, L. (1998). The productive and optical status of the oligotrophic waters of the Southern Aegean Sea (Cretan Sea), Eastern Mediterranean. Journal of Plankton Research, 20(5): 985–995.

Ignatiades, L. (2002a). Statistical applications in marine phytoplanktonic studies: Part I. Parametric and non-parametric statistical analysis. pp. 49–57. *In*: Subba Rao, D.V. (Ed.). Pelagic Ecology Methodology. A.A. Balkema Publishers. Lisse. 464pp.

Ignatiades, L. (2002b). Statistical applications in marine phytoplanktonic studies: Part II. Analysis of ecological data. pp. 59–65. *In*: Subba Rao, D.V., (Ed.). Pelagic Ecology Methodology. A.A. Balkema Publishers. Lisse. 464pp.

Ignatiades, L. (2005). Scaling of the trophic status of the Aegean Sea, Eastern Mediterranean. Journal of the Sea Research, 54: 51–57.

Ignatiades, L., Gotsis-Skretas, O. and Metaxatos, A. (2007). Field and culture studies on the ecophysiology of the toxic dinoflagellate *Alexandrium minutum* (Halim) present in Greek coastal waters. Harmful Algae, 6(2): 153–165.

Ignatiades, L., Gotsis-Skretas, O., Pagou, P. and Krasakopoulou, E. (2009). Diversification of phytoplankton community structure and related parameters along a large scale longitudinal east-west transect of the Mediterranean Sea. Journal of Plankton Research, 31(4): 411–428.

Ignatiades, L. and Gotsis-Skretas, O. (2010). A review on toxic and harmful algae in Greek coastal waters (E. Mediterranean Sea). Toxins, 2: 1019–1037.

IOC-UNESCO. (2009). Annual Report. 10C Annual Reports Series, No. 16 UNESCO.

IPCC. (2007). A report of working group 1 of the Intergovernmental Panel on Climate Change. Summary of Policy Makers and Technical Summary.

Istvanovics, V., Osztoics, A. and Honti, M. (2004). Dynamics and ecological significance of daily internal load of phosphorus in shallow Lake Balaton, Hungary. Freshwater Biology, 49(3): 232–252.

Ivanov, V.A. and Belokopytov, V.N. (2013). Oceanography of the Black Sea. ECOZY-Gidrofizika, Sevastopol, 210pp.

Iverson, R.L., Esaias, W.E. and Turpie, K. (2000). Ocean annual phytoplankton carbon and new production, and annual export production estimated with empirical equations and CZCS data. Global Change Biology, 6: 57–72.

Jaccard, J., Becker, M.A. and Wood, G. (1984). Pairwise multiple comparison procedures—a review. Phycological Bulletin, 96: 589–596.

Jacobson, M.A. (1995). The United Nations' Regional Seas Programme Seas Programme: how does it measure up? Coastal Management, 23: 19–39.

Jafari, N. (2010). Review on pollution sources and control in the Caspian Sea region. Journal of Ecology of the Natural Environment, 2: 25–29.

Janssen, R. (1992). Multiobjective Decision Support for Environmental Management. Kluwer Publishers. Amsterdam.

Janssen, R. (2001). On the use of multi-criteria analysis in environmental impact assessment in The Netherlands. Journal of Multi-Criteria Decision Analysis, 10: 101–109.

Jasanoff, S. (1990). The Fifth Branch. Science Advisors as Policy Makers. Harvard University Press, Cambridge MA, London.

Jessen, C., Roder, C., Lizcano, J.V., Voolstra, C.R. and Wild, C. (2012). Top-down and bottom-up effects on Red Sea coral reef algae. Proceedings of the 12th International Coral Reef Symposium, Cairnns, Australia, 9–13 July 2012. 11A Ecology and Macroecology: general sessions.

Jiang, Q., He, J., Wu, J., Hu, X., Ye, G. and Christakos, G. (2018). Assessing the severe eutrophication status and spatial trend in the coastal waters of Zhejiang province (China). Limnology and Oceanography, 9999: 1–15.

Jickells, T., An, Z.S. and Andersen, K.K. (2005). Global iron connections between desert dust, ocean biochemistry and climate. Science, 308: 67–71.

Johannesson, K. and Andre, C. (2006). Life on the margin: genetic isolation and diversity loss in a peripheral marine ecosystem, the Baltic Sea, Molecular Ecology, 15: 2013–2029.

John, A.W.G., Reid, P.C., Batten, S.D. and Anang, E.R. (2002). Monitoring levels of phytoplankton color in the gulf of guinea using ships of opportunity. pp. 141–146. *In*: McGlade, J.M., Cury, P., Koranteng, K.A. and Hardman-Mountford, N.J. (Eds.). The Gulf of Guinea: Large Marine Ecosystem: Environmental Forcing and Sustainable Development of Marine Resources. Elsevier. Amsterdam.

Jones, R.A. and Lee, G.F. (1982). Recent advances in assessing impact of phosphorus loads on eutrophication-related water quality. Journal of Water Research, 16: 503–515.

Kaniaru, D. (2000). Development and implementation of Environmental Law—a contribution to UNEP. Environmental Policy and Law, 30: 234.

Karadanelli, M., Moriki, A., Faridis, E. and Karydis, M. (1992). Annual pattern of heterotrophic bacteria and phytoplankton in a nutrient rich coastal system. Rapp. Comm. Int. Mer. Medit. 33: 198.

Karl, D.M., Christian, J.R., Dore, S.E. et al. (1999). Seasonal and interannual variability in primary production and particle flux at Station ALOHA. Deep Sea Research. Part II top. Stud. Oceanogra., 43: 239–268.

Karpinsky, M.G., Katunin, D.N., Goryunova, V.B. and Shiganova, T.A. (2005). Biological features and resources. pp. 191–210. *In*: Kostianoy, A. and Kosarev. A. (Eds.). The Caspian Sea Environment. Springer.

Karydis, M. (1981). The toxicity of crude oil for the marine algal *Skeletonema costatum* (Greville) Cleve in relation to nutrient limitation. Hydrobiologia, 85: 137–143.

Karydis, M., Ignatiades, L. and Moschopoulou, N. (1983). An index associated with nutrient eutrophication in the marine environment. Estuarine, Coastal and Shelf Science, 16: 339–344.

Karydis, M. (1992). Scaling methods in assessing environmental quality—a methodological approach to eutrophication. Environmental Monitoring and Assessment, 22: 123–136.

Karydis, M. (1994). Environmental quality assessment based on the analysis of extreme values—a practical approach for evaluating eutrophication. Journal of Environmental Science and Health. A: Environmental Science and Engineering, 29: 775–791.

Karydis, M. (1996). Quantitative assessment of eutrophication: a scoring system for characterizing water quality in coastal marine ecosystems. Environmental Monitoring and Assessment, 41: 233–246.

Karydis, M. and Tsirtsis, G. (1996). Ecological indices: a biometric approach for assessing eutrophication levels in the marine environment. The Science of the Total Environment, 186: 209–219.

Karydis, M. and Tsirtsis, G. (1999). Application of the discriminant analysis for assessing coastal marine eutrophication. International Conference of Science and Technology, Athens, Greece, pp. 243–249 (in Greek, English Abstract).

Karydis, M. (2009). Eutrophication assessment of coastal waters based on indicators: a literature review. Global Nest Journal, 11: 373–390.

Karydis, M. and Kitsiou, D. (2012). Eutrophication and environmental policy in the mediterranean sea: a review. Environmental Monitoring and Assessment, 184: 4931–4984.

Karydis, M. and Kitsiou, D. (2013). Marine water quality monitoring: a review. Marine Pollution Bulletin, 77: 23–36.

Karydis, M. (2013). Public attitudes and environmental impacts of wind farms: a review. Global NEST Journal, 15(4): 581–600.

Karydis, M. (2014). Use of aquatic microcosm systems in phytoplankton ecology studies: objectives, limitations and applications. pp. 245–275. *In*: Sebastia, M.T. (Ed.). Phytoplankton: Biology, Classification and Environmental Impacts. Nova Science Publishers, Inc. NY. 348pp.

Karydis, M. and Kitsiou, D. (2014). Eutrophication in the European Regional Seas: a review on impacts, assessment and policy. pp. 167–243. *In*: Sebastia, M.T. (Ed.). Phytoplankton: Biology, Classification and Environmental Impacts. Nova Science Publishers, Inc. NY. 348pp.

Karydis, M. (2015). Environmental marine monitoring strategies and coastal ecosystem management: matching science with policy. pp. 13–41. *In*: Sebastia, M.T. (Ed.). Coastal Ecosystems: Experiences and Recommendations for Environmental Monitoring Programs. Nova Science Publishers. NY. 220pp.

Karydis, M. (2017). Water quality and ecosystems' health in Oceans and Seas around the World. pp. 3–36. *In*: Kitsiou, D. and Karydis, M. (Eds.). Marine Spatial Planning: Methodologies, Environmental Issues and Current Trends. Nova Science Publishers, Inc. NY. 483pp.

Kautsky, N., Kautsky, H., Kautsky, U. and Waren, M. (1986). Decreased depth penetration of *Fucus vesiculosus* (L.) since the 1940's indicates eutrophication in the Baltic Sea. Marine Ecology Progress Series, 28: 1–8.

Keckes, S. (1994). The regional seas programme: integrating environment and development: the nest phase. *In*: Payoyo, P.B. (Ed.). Ocean Governance: Sustainable Development of the Seas. United Nations, New York.

Keohane, R.O. and Levy, M. (Eds.). (1996). Institutions for Environmental Aid: Pitfalls and Promise. MIT Press, Cambridge, Massachusetts.

Kersten, M., Dicke, M., Kriews, M. et al. (1988). Distribution and fate of heavy metals in the North Sea. pp. 300–347. *In*: Salomons, W., Bayne, B.L., Duursma, E.K. and Forstrner, U. (Eds.). Pollution of the North Sea: An Assessment. Springer-Verlag.

Ketchum, B.H. (1969). Eutrophication of estuaries. pp. 197–209. *In*: Eutrophication, Causes, Consequences, Correctives. National Academy of Sciences, Washington.

Ketchum, B.H. (1983). Estuarine characteristics. pp. 1–14. *In*: Ketchum, B.H. (Eds.). Estuaries and Enclosed Seas, Ecosystems of the World. Elsevier NY.

Kidd, S., Plater, A. and Frid, C. (Eds.) (2011). The Ecosystem Approach to Marine Planning and Management. Earthscan. London. 231pp.

Kideys, A.E., Kovalev, A.V., Shulaman, G., Gordina, A. and Bingel, F. (2000). A review of zooplankton investigations of the Black Sea over the last decade. Journal of Marine Systems, 24: 355–371.

Kideys, A.E. (2002). Fall and rise of the Black Sea ecosystem. Science, 297: 1482–1484.

King, M. (2004). From Mangroves to Coral Reef: Sea Life and Marine Environments in Pacific Islands. SPREP, Available at: www.sprep.org.ws.

Kirchner, W.B. and Dillon, P.J. (1975). An empirical method of estimating the retention of phosphorus in lakes. Water Resource Research, 11(1): 182–183.

Kitsiou, D. and Karydis, M. (1998). Development of categorical mapping for quantitative assessment of eutrophication. J. Coastal Conservation, 4: 35–44.

Kitsiou, D. and Karydis, M. (2000). Categorical mapping of marine eutrophication based on ecological indices. The Science of the Total Environment, 255: 113–127.

Kitsiou, D., Tsirtsis, G. and Karydis, M. (2001). Developing an optimal sampling design. A case study in a coastal marine ecosystem. Environmental Monitoring and Assessment, 71: 1–12.

Kitsiou, D., Coccossis, H. and Karydis, M. (2002). Multi-dimensional evaluation and ranking of coastal areas using GIS and multiple criteria choice methods. The Science of the Total Environment, 284: 1–17.

Kitsiou, D. and Karydis, M. (2011). Coastal marine eutrophication assessment: a review on data analysis. Environment International, 37: 778–801.

Kitsiou, D., Kostopoulou, M., Karydis, M. and Tsirtsis, G. (2011). Assessment of the environmental quality in Strymonikos Gulf, North Aegean Sea, Greece. Technical report, Univ. of the Aegean, Dept. of Marine Sciences (in greek).

Kitsiou, D. and Karydis, M. (2014). Oil spills: behavior of oil, impact, detection, tracking and management. pp. 197–242. *In*: Jefferson, D.E. (Ed.). Marine Pollution: Types, Environmental Significance and Management Strategies. Nova Science Publ. New York.

Kitsiou, D. and Karydis, M. (Eds.). (2017). Marine Spatial Planning: Methodologies, Environmental Issues and Current Trends. Nova Science Publishers. 483pp.

Kitsiou, D., Politi, E. and Kostopoulou, M. (2017). The application of zoning and multiple criteria analysis in marine spatial planning. pp. 175–195. *In*: Kitsiou, D. and Karydis, M. (Eds.). Marine Spatial Planning: Methodologies, Environmental Issues and Current Trends. Nova Science Publishers. 483pp.

Klemans, V. (2012). Remote sensing of algal blooms: an overview with case studies. Journal of Coastal Research, 28(1A): 34–43.

Koetse, M.J., Brouwer, R. and van Beukering, J.H. (2015). Economic valuation methods for ecosystem services. pp. 108–131. *In*: Bouma, J.A. and van Beukering, P.J.H. (Eds.). Ecosystem Services: from Concept of Practice. Cambridge University Press.

Kong, X., Sun, Y., Su, R. and Shi, X. (2017). Real-time eutrophication status evaluation of coastal waters using support vector machine with grid search algorithm. Marine Pollution Bulletin, 119(1): 307–319.

Konovalov, S.K., Ivanov, L.I., Murray, J.W. and Eremeeva, L.V. 1999. Eutrophication: a plausible cause for changes in the hydrochemical structure of the Black Sea anoxic layer. *In*: Besiktepe, S.T., Unluata, U. and Bologa, A.S. (Eds.). Environmental Degradation of the Black Sea: Challenges and Remedies. NATO-ASI, 56: 61–74 (Berlin).

Konovalov, S.K. and Murray, J.W. (2001). Variations in the chemistry of the Black Sea on a time scale of decades (1960–1995). Journal of Marine Systems, 31: 217–243.

Konovalov, S.M. (1995). Anthropogenic impact and ecosystems of the Black Sea. Bulletin de L' Institute Oceanographique, Monaco, 15: 53–83.

Koo, B.K. and O'Connell, P.E. (2006). An integrated modelling and multicriteria analysis approach to managing nitrate diffuse pollution: 1. Framework and methodology. The Science of the Total Environment, 359: 1–16.

Korshenko, A. and Gul, A.K. (2005). Pollution of the Caspian Sea. pp. 109–142. *In*: Kostianoy, A. and Kosarev, A. (Eds.). The Caspian Sea Environment. Springer.

Kosme, N., Koski, M. and Hauschild, M.Z. (2015). Exposusre factors for marine eutrophication impacts assessment based on a mechanistic biological model. Ecological Modelling, 317: 50–63.

Kouassi, A.M., Kab, N. and Metongo, S. (1995). Land-based sources of pollution and environmental quality of the Ebrie lagoon waters. Marine Pollution Bulletin, 30: 295–300.

Kouassi, A.M. and Biney, C. (1999). Overview of the marine environmental problems of the West and Central African Region. Ocean and Coastal Management, 42: 71–76.

Kovač, Z., Morović, M. and Matić, M. (2014). Uncovering spatial and temporal patterns of Adriatic Sea colour with self-organizing maps. International Journal of Remote Sensing, 35(6): 2105–2117.

Kovacs, A. and Honti, M. (2008). Estimation of diffuse phosphorus emissions at small catchment scale by GIS-based pollution potential analysis. Desalination, 226: 72–80.

Kozarev, A.N. (2005). Physico-geographical conditions of the Caspian Sea. pp. 5–32. *In*: Kostianoy, A. and Kosarev, A. (Eds.). The Caspian Sea Environment. Springer.

Kozarev, A.N. and Kostianov, A. (Eds.). (2005). The Caspian Sea Environment. Springer.

Krebs, C.J. (1999). Ecological Methodology. Addison-Wesley Longman Publishers.

Krishnamurthy, A., Mooze, J.K., Zender, C.S. and Luo, C. (2007). Effects of atmospheric inorganic nitrogen deposition on ocean biochemistry. Journal of Geophysical Research, 112: G02019. http://dx.doi.org/10.029/2006JG000334.

Krom, M., Hebut, B. and Mantoura, R. (2004). Nutrient budget for the Eastern Mediterranean: implications for phosphorus limitation. Limnology and Oceanography, 49: 1582v1592.

Krom, M.D., Kress, N., Brenner, S. and Gordon, L.I. (1991). Phosphorus limitation of primary productivity in the Eastern Mediterranean Sea. Limnology and Oceanography, 36: 424–432.

Krom, M.D., Emeis, K.C. and Van Cappellen, P. (2010). Why is the Mediterranean phosphorus limited? Progress in Oceanography, 85: 236–244.

Kyewalyanka, M., Naik, R., Hegde, S., Raman, M., Barlow, R. and Roberts, M. (2007). Phytoplankton biomass and primary production in Delagoa Bight Mozambique: Application of remote sensing. Estuarine Coastal and Shelf Science, 74: 429–436.

Kyewalyanka, M. (2015). Phytoplankton and primary production. pp. 213–232. *In*: UNEP-Nairobi Convention and WIOSMA: The Regional State of the Coast Report: Western Indian Ocean. UNEP and WIOSMA, Nairobi, Kenya, 546pp.

Kyvelou, S.S. and Pothitaki, I.V. (2017). Current attitudes and lessons learnt in maritime/marine spatial planning. pp. 71–92. *In*: Kitsiou, D. and Karydis, M. (Eds.). Marine Spatial Planning: Methodologies, Environmental Issues and Current Trends. Nova Science Publishers. 483pp.

Lam, C.W.Y. and Ho, K.C. (1989). Red tides in Tolo Harbour, Hong Kong. pp. 49–52. *In*: Okaichi, T., Anderson, D.M. and Nemoto, T. (Eds.). Red Tides: Biology, Environmental Science and Toxocology, New York, Elsevier Science.

Lambshead, P.J.D., Platt, H.M. and Shaw, K.M. (1983). The detection of differences among assemblages of marine benthic species on an assessment of dominance diversity. Journal of Natural History, 17(6): 859–874.

Lampert, L., Queguiner, B., Labasque, T., Pichon, A. and Lebreton, N. (2002). Spatial variability of phytoplankton composition and biomass on the eastern continental shelf of the Bay of Biscay (north-east Atlantic Ocean). Evidence for a bloom of Emiliania huxleyi (Prymnesiophyceae) in spring 1998. Cont. Shelf Res., 22: 1225–1247.

Lancaster, P. and Salkauskas, K. (1986). Curve and Surface Fitting: An Introduction. London: Academic Press.

Lancelot, C., Billen, G., Sournia, A. et al. (1987). Phaeocystis blooms and nutrient enrichment in the continental zones of the North Sea. Ambio, 16: 38–46.

Larsson, U., Elmgren, R. and Wulff, F. (1985). Eutrophication and the Baltic Sea—causes and consequences. AMBIO, 14: 9–14.

Lassus, P., Herbland, A. and Lebaut, C. (1991). Dinophysis blooms and toxic effects along the French Coast. World Aquaculture, 22(4): 49–54.

Lattemann, S. and Hopner, T. (2008). Environmental impact and impact assessment of seawater desalination. Desalination, 220: 1–15.

Lazzari, P., Solidoro, C. and Ibello, V. (2011). Seasonal and inter-annual variability of plankton chlorophyll and primary production in the Mediterranean Sea: a modelling approach. Biogeoscience Discussions, 8: 5379–5422.

Lee, A.J. (1980). North Sea: physical oceanography. pp. 467–493. In: Banner, F.T., Collins, M.B. and Massie, K.S. (Eds.). The North West European Shelf Seas: The Seabed and the Sea in Motion. II. Physical and Chemical Oceanography and Physical Resources. Elsevier, Amsterdam.

Lefevre, D., Minas, H.J., Minas, M., Robinson, C., LeB, W.P.J. and Woodward, E.M.S. (1997). Review of gross community production, primary production, net community production and dark community respiration in the Gulf of Lions. Deep Water Research II, 44(3-4): 801–832.

Legendre, P. (1993). Spatial autocorrelation-trouble or new paradigm. Ecology, 74(6): 1659–1673.

Legendre, P. and Legendre, L. (2003). Numerical Ecology: Developments in Environmental Modelling. 2nd Edition. Elsevier, Amsterdam. 853pp.

Leliaert, F. and Coppejans, E. (2004). Seagrasses and seaweeds. In: Standard Survey Methods for Key Habitats and Key species in the Red Sea and Gulf of Aden. PERSGA Technical Series, No. 10, 101–124, PERSGA, Heddah, 310pp.

Lelong, A., Hegaret, H. and Soudant, P. (2014). Link between domoic acid production and cell physiology after exchange of bacterial communities between toxic Pseudo-nitzschia multispecies and non-toxic Pseudo-nitzschia delicatissima. Marine Drugs, 12: 3587–3607.

Lenes, J.M., Darrow, B.P., Cattrall, C. et al. (2001). Iron fertilization and response on the West Florida shelf. Limnology and Oceanography, 46: 1261–1277.

de Leo, G.A., Bartoli, M., Naldi, M. and Viaroli, P. (2002). A first generation stochastic bioeconomic analysis of algal bloom control in a coastal lagoon (Sacca id Goro, Po River Delta). Marine Ecology, 23: 92–100.

Li, L. and Revesz, P. (2004). Interpolation methods for spatio-temporal geographic data. Computers, Environment and Urban Systems, 28: 201–227.

Liu, H., Zhou, Y. and Xiao, W. et al. (2012). Shifting nutrient mediated interactions between algae and bacteria in a microcosm: evidence from alkaline phosphatase assay. Microbiological Research, 167: 292–298.

Liu, S.M., Hong, G.H., Zhang, J., Ye, W. and Jiang, X.L. (2009). Nutrient budgets for large Chinese estuaries. Biogeosciences, 6: 2245–2263.

Loaiza, I., Hurtado, D., Miglio, M. et al. (2015). Tissue specific Cd and Pb accumulation in Peruvian scallop (Argopecten purpuratus) transplanted to a suspended and bottom culture at Sechura Bay, Peru. Marine Pollution Bulletin, 91: 429–440.

Loehle, C. (1983). Evaluation of theories and calculation tools in ecology. Ecological Modelling, 19: 239–247.

Lourdes, M., McGlone, S.D., Azanza, R.V., Villanoy, C.I. and Jacinto, G.S. (2008). Eutrophic waters, algal boom and fish kill in fish farming area in Bolindo, Pangasinan, Philippines. Marine Pollution Bulletin, 57: 295–301.

Loya, Y., Lubinevsky, H., Rosenfeld, M. and Kramarsky-Winter, E. (2005). Nutrient enrichment caused by in situ fish farms at Eilat, Red Sea is detrimental to coral reproduction. Marine Pollution Bulletin, 49: 344–353.

Lu, F., Chen, Z. and Liu, W. (2014). A GIS-based system for assessing marine water quality around offshore platforms. Ocean & Coastal Management, 102(A): 294–306.

Luck, G.W. et al. (2009). Quantifying the contribution of organisms to the provision of ecosystem services. BioScience, 59: 223–235.

Ludwig, J.A. and Reynolds, J.F. (1988). Statistical Ecology: A Primer on Methods and Computing. J. Wiley and Sons. New York.

Ludwig, W., Dumont, E., Meybeck, M. and Heussner, S. (2009). River discharges of water and nutrients to the Mediterranean and Black Sea: major drivers for ecosystem changes during past and future decades? Prog. Oceanogr., 80(3): 199–217.

Lugendo, B. (2015). Mangroves, salt marshes and seagrass. *In*: UNEP-Nairobi Convention and WIOSMA: The Regional State of the Coast Report: Western Indian Ocean. UNEP and WIOSMA, Nairobi, Kenya, 546pp.

Lundberg, C., Lonnoroth, M., VonNumers, M. and Bondsdorff, E. (2005). A multivariate assessment of coastal eutrophication. Examples from the Gulf of Finland, Northern Baltic Sea. Marine Pollution Bulletin, 50: 1185–1196.

Lyons, D.A., Arvanitidis, C., Blight, A.J. et al. (2014). Macroalgal blooms alter community structure and primary productivity in marine communities. Global Change Biology, 20: 2712–2724.

Mabit, L. and Bernard, C. (2007). Assessment of spatial distribution of fallout radionuclides through geostatistics concept. Journal of Environmental Radioactivity, 97: 206–219.

MacArthur, R.H. (1972). Geographical Ecology: Patterns in the Distribution of Species. Princeton University Press, Princeton, NJ.

Magurran, A.E. (2004). Measuring Biological Diversity. Blackwell Publishing, Oxford. 256pp.

Mahongo, S. and Mwaipopo, R. (2015). Planet and life. *In*: UNEP-Nairobi Convention and WIOSMA: The Regional State of the Coast Report: Western Indian Ocean. UNEP and WIOSMA, Nairobi, Kenya, 546pp.

Mahowald, N., Jickells, T.D., Tomothy, D. et al. (2008). The global distribution of atmospheric phosphorus sources, concentrations and deposition rates and anthropogenic impacts. Global Biogeochemical Cycles, 22: GB4026.

Maina, J. (2015). Beaches and the nearshore. pp. 41–52. *In*: UNEP-Nairobi Convention and WIOSMA: The Regional State of the Coast Report: Western Indian Ocean. UNEP and WIOSMA, Nairobi, Kenya, 546pp.

Malczewski, J. (2006). GIS-based multicriteria decision analysis: a survey of the literature. International Journal of Geographical Information Science, 20: 703–726.

Manly, B.F.J. (2001). Statistics for Environmental Science and Management. Chapman and Hall. CRC Press.

Mantua, N. (2004). Methods for detecting regime shifts in large marine ecosystems: a review with approaches applied to North Pacific data. Progress in Oceanography, 60: 165–182.

MAP. (2008). Municipal Wastewater Treatment plants in Mediterranean coastal cities. MAP Technical Reports, Series No. 109.

Marchetti, R. (1985). Indagini sul problema dell' Emilia-Romagna. Ed. Regione Emilia-Romagna Assessorato Ambiente e Difesa de sualo Bologna. 1–308pp.

Marchuk, G.L., Kordzadze, A.A. and Skiba, Y.N. (1975). Calculation of the basic hydrological fields in the Black Sea. Atmospheric Ocean Physics, 11: 379–393.

Marcotullio, P.J. (2001). Asian urban sustainability in the era of globalization. Habitat International, 25: 577–598.

Maritorena, S. and Siegel, D.A. (2005). Consistent merging of satellite ocean color data sets using a bio-optical model. Remote Sensing of Environment, 94: 429–440.

Marrugo-Gonzalez, A.J., Fernandez-Maestre, R. and Alm, A.A. (2000). The pacific Coast of Colombia. Chapter 43. pp. 677–686. *In*: Sheppard, Ch. (Ed.). Seas at the Millenium Environmental Evaluation, Vol. 1. Pergamon Press, Amsterdam.

Matheron, G. (1963). Principles of geostatistics. Economic Geology, 58: 1246–1266.

McGlade, J.M., Cury, P., Koranteng, K.A. and Hardman-Mountford, N.J. (Eds.). (2002). The Gulf of Guinea: Large Marine Ecosystem: environmental forcing and sustainable development of marine resources. Elsevier. Amsterdam.

McLeod, E., Chmura, G. and Bouillon, S. et al. (2011). A blueprint for blue carbon: toward an improved understanding of the role of vegetated coastal habitats in sequestering CO_2. Frontiers in Ecology and the Environment, 9: 552–560.

McRoy, C.P. and Goering, J.J. (1974). Nutrient transfer between the seagrass *Zostera marina* and its epiphytes. Nature, 248: 173–174.

Ménesguen, A. and Lacroix, G. (2018). Modelling the marine eutrophication: A review. Science of the Total Environment, 636: 339–354.

Michelakaki, M. and Kitsiou, D. (2005). Estimation of anisotropies in chlorophyll*a* spatial distributions based on satellite data and variography. Global NEST Journal, 7(2): 204–211.

Mikaelyan, A.S., Pautova, L.A., Chasovnikov, V.K., Mosharov, S.A. and Silkin, V.A. 2015. Alternation of diatoms and coccolithophores in the north-eastern Black Sea: a response to nutrient changes. Hydrobiologia 755(1): 89–105.

Mikaelyan, A.S., Chasovnikov, V.K., Kubryakov, A.A. and Stanichny, S.V. (2017). Phenology of drivers of the winter-spring phytoplankton bloom in the open Black Sea: the application of Svedrup's hypothesis and its refinements. Progress in Oceanography, 151: 163–176.

Miles, E.L. and Underdal, A. (Eds.). (2001). Explaining Environmental Regime Effectiveness: Confronting Theory with Evidence, MIT Press, Cambridge, Massachusetts.

Millenium Ecosystem Assessment. (2005). Ecosystems and Human Well-being. General Synthesis. Island Press, Washington D.C.

Mingazzini, M., Rinaldi, A. and Montanari, G. (1992). Multilevel nutrient enrichment bioassays on North Adriatic coastal waters. pp. 365–369. *In*: Vollenweider, R.A., Marchetti, R. and Viviani, R. (Eds.). Marine Coastal Eutrophication. Proceedings of an International Conference, Bologna, Italy, 21–24 March 1990. Science of the Total Environment. Elsevier. Region 49, UNEP & GEF, Kalmar University.

Mistafa, N.T. and Ali, O.M.M. (2005). The Red Sea and Gulf of Aden: thematic report of the GIWA.

MIT. (1970). Man's impact on the global environment. Report of the Study of Critical Environment Problems (SCEP): Assessments and Recommendations for Action. Massachusetts Institute of Technology, MIT Press. Cambridge, Mass. and London.

Mmochi, A.J. and Mwandya, A.W. (2003). Water quality in the Integrated Mariculture Pond Systems (IMPS) at Makoba Bay, Zanzibar, Tanzania. Western Indian Ocean Journal of Marine Science, 2: 15–24.

Moisander, P.H., McClinton, E. and Paerl, H.W. (2002). Salinity effects on growth, photosynthetic parameters and nitrogenize activity in estuarine planktonic cyanobacteria. Microbial Ecology, 43: 432–442.

Momanyi, A. (2015). Governance: legal and institutional frameworks. pp. 445–458. *In*: UNEP-Nairobi Convention and WIOSMA: The Regional State of the Coast Report: Western Indian Ocean. UNEP and WIOSMA, Nairobi, Kenya, 546pp.

Moncheva, S., Gotsis-Skretas, O., Pagou, K. and Krastev, A. (2001). Phytoplantkon blooms in Black Sea and Mediterranean coastal ecosystems subjected to anthropogenic eutrophication: similarities and differences. Estuarine, Coastal and Shelf Science, 53(3): 281–295.

Monod, J. (1942). Recherches sur la croissance des cultures bacteriennes. Herman. Paris. 210pp.

Moon, J.Y., Lee, K., Tanhua, T., Kress, N. and Kim, I.N. (2016). Temporal nutrient dynamics in the Mediterranean Sea in response to anthropogenic inputs. Geophysical Research Letters, 10.1002/2016GL068788, 5243–5251.

Moradi, M. and Kabiri, K. (2015). Spatio-temporal variability of SST and Chlorophyll-α from MODIS data in the Persian Gulf. Marine Pollution Bulletin, 98(1-2): 14–25.

Morales-Baquero, R., Pullido-Villena, E. and Peche, I. (2013). Chemical signature of Saharan dust on dry and wet atmospheric deposition in the south-western Mediterranean region. Tellus, Ser. B, 65: 18720.

Morelli, B., Hawkins, T.R., Niblick, B., Henderson, A.D., Golden, H.E., Compton, J.E., Cooter, E.J. and Bare, J.C. (2018). Critical review of eutrophication models for life cycle assessment. Environmental Science and Technology, 52: 9562–9578.

Moriki, A. and Karydis, M. (1994). Application of multicriteria choice methods in assessing eutrophication. Environmental Monitoring and Assessment, 33: 1–18.

Morse, D., Tse, S.P.K and Lo, S.C.L. (2018). Exploring dinoflagellate biology with high-throughput proteomics. Harmful Algae, 75: 16–26.

Morton, S.L., Vershinin, A., Smith, L.L. et al. (2009). Seasonality of *Dinophysis* spp. and *Prorocentrum lima* in Black Sea phytoplankton and associated shellfish toxicity. Harmful Algae, 8: 629–636.

Moskalenko, L.V. (1976). Calculation of stationary wind driven currents in the Black Sea. Oceanology, 15: 168–171.

Mousing, E.A., Adjou, M. and Ellegocard, M. (2015). Evidence of intensified silica recycling in the Black Sea after 1970. Estuarine, Coastal and Shelf Science, 164: 335–339.

Mu, L. (2009). Thiessen polygon. International Encyclopedia of Human Geography, 231–236.

Nadim, F., Bagtzoglou, A.C. and Iranmahboob, J. (2008). Coastal management in the Persian Gulf region within the framework of the ROPME programme of action. Ocean and Coastal Management, 51: 556–565.

NAS. (1981). Committee to review methods for ecotoxicology testing for effects of chemicals on ecosystems. National Academy of Sciences, National Academy Press, Washington, D.C.

Nasrollahzadeh, H.S., Bin Din, Z. and Foong, S.Y. (2008). Trophic status of the Iranian Caspian Sea based on water quality parameters and phytoplankton diversity. Continental Shelf Research, 28(9): 1153–1165.

National Academy of Sciences. (2003). Clean Coastal Waters: Understanding and Reducing the Effects of Nutrient Pollution. Washington: National Academy Press.

Naumann, M.S., Bednarz, V.N., Ferse, S.C.A., Nigl, W. and Wild, C. (2015). Monitoring of coastal reefs near Dahab (Gulf of Aqaba, Red Sea) indicates local eutrophication as potential cause for change in benthic communities. Environmental Monitoring and Assessment, 187: 44.

Newton, A., Icely, J.D., Falcao, M., Nobre, A., Nunes, J.P., Ferreira, J.G. and Vale, C. (2003). Evaluation of eutrophication in the Ria Formosa coastal lagoon, Portugal. Continental and Shelf Research, 23: 1945–1961.

Nezlin, N.P. (2005). Patterns of seasonal and interannual variability of remotely sensed chlorophyll. pp. 143–158. *In*: Kostianoy, A. and Kosarev, A. (Eds.). The Caspian Sea Environment. Springer.

Nielsen, S.E., Bayne, E.M., Schieck, J., Herbers, J. and Boutin, S. (2007). A new method to estimate species and biodiversity intactness using empirically derived reference conditions. Biological Conservation, 137: 403–414.

Nijkamp, P. and Voogd, H. (1986). A survey of qualitative multiple criteria choice models. *In*: Nijkamp, P., Leitner, H., Wringley, N. (Eds.). Measuring the Unmeasurable. Dordrtecht: Kluwer Nijhoff, 425–447.

Nikolaidis, G., Patoucheas, D.P. and Moschandreou, K. (2006). Estimating breakpoints of chlα in relation with nutrients from Thermaikos Gulf (Greece) using piecewise linear regression. Fresenius Environmental Bulletin, 15(9B): 1189–1192.

Nixon, S. (1990). Marine eutrophication: a growing international problem. Ambio, 19: 101.

Nixon, S.W. (1995). Coastal marine eutrophication: a definition, social causes and future concerns. Ophelia, 41: 199–219.

Nixon, S.W. (2009). Eutrophication and the macroscope. Hydrobiologia, 629: 5–19.

NOOA. (2011). Large Marine Ecosystems of the World. Retrieved from: www.lme.noaa.gov.

NRC. (1990). Managing Troubled Waters. The Role of Marine Environmental Monitoring. National Academy of Sciences, National Academy Press, Washington, D.C.

NRC. (1999). New Strategies for American Watersheds. National Academy Council, National Academy Press, Washington, DC.

NRC. (2000). Clean Coastal Waters: Understanding and Reducing the Effects of Nutrient Pollution. National Academy Press.

NSFT. (1994). The 1993 Quality Status Report of the North Sea. Ducrotoy, J.P. et al. (Eds.). North Sea Task Force, Oslo and Paris Commissions/International Council of the Exploration of the Sea. Olsen and Olsen.

Nunneri, C. and Hofmann, J. (2005). A participatory approach for Integrated River Basin Management in the Elbe catchment. Estuarine, Coastal and Shelf Science, 62: 521–537.

O'Higgins, T.G. and Gilbert, A.J. (2014). Embedding ecosystem services into the Marine Strategy Framework Directive: illustrated by eutrophication in the North Sea. Estuarine, Coastal and Shelf Science, 140: 146–152.

O'Neil, J.M., Davis, T.W., Burford, M.A. and Gobler, C.J. (2012). The use of harmful cyanobacteria blooms: the potential roles of eutrophication and climate change. Harmful Algae, 14: 313–334.

O'Sullivan, P.E. (1995). Eutrophication. International Journal of Environmental Studies, 47: 173–193.

Obura, D. (2015). Coral and biogenic reef habitats. pp. 71–84. *In*: UNEP-Nairobi Convention and WIOSMA: The Regional State of the Coast Report: Western Indian Ocean. UNEP and WIOSMA, Nairobi, Kenya, 546pp.

Obura, D.O., Church, J.E. and Gabrie, C. (2012). Assessing marine world heritage from an ecosystem perspective: the Western Indian Ocean. World Heritage Center, United Nations Education, Science and Cultural Organization (UNESCO).

OCTA. (2015). Overseas Countries and Territories: Environmental Profiles Europe Aid/127054/C/ SR/multi. Final Report Section D: Pacific Region.

OECD. (1984). Coastal Area Management and Development. United Nations Department of International Economic and Social Affairs. Ocean Economics and Technology Branch. Prergamon Press. 188pp.

OECD. (1996). DAC Guidelines on Aid and Environment: No. 8. Guidelines for Aid Agencies on Global and Regional Aspects of the Development and Protection on the Marine and Coastal Environment. Development Co-operation Directorate: OECD.

Oerder, V., Colas, F., Echevin, V. et al. (2015). Peru-Chile upwelling dynamics under climate change. Journal of Geophysical Research: Oceans. 120: 1152–1172, doi:10.1002/2014JC010299.

Officer, C.B. and Ryther, J.H. (1980). The possible importance of silicon in marine eutrophication. Marine Ecology Progress Series, 3: 83–91.

Ogata, T., Pholpunthin, P., Fuluyo, Y. and Kodama, M. (1990). Occurrence of Alexandrium cohorticula in Japanese coastal water. Journal of Applied Phycology, 2: 351–356.

Oguz, T. (2005). Long term impacts of anthropogenic forcing on the Black Sea Ecosystem. Oceanography, I, 18: 112–121.

Okaichi, T. (1989). Red tide problems in the Seto Island, Japan. pp. 137–142. *In*: Okaichi, T., Anderson, D.M. and Nemoto, T. (Eds.). Red Tides: Biology, Environmental Science and Toxocology, New York, Elsevier Science.

Okin, G.S., Baker, A.R. and Tegen, I. (2011). Impacts of atmospheric nutrient deposition on marine productivity: roles of nitrogen, phosphorus and iron. Global Biogeochemical Cycles, 25: GB2022.

Okolodkov, Y.B. and Dodge, J.D. (1996). Biodiversity and biogeography of planktonic dinoflagellates in the Arctic Ocean. Journal of Experimental Marine Biology and Ecology, 202: 19–27.

Orfanidis, S., Panagiotidis, P. and Stamatis, N. (2001). Ecological evaluation of transitional and coastal waters: a marine benthic macrophytes based model. Mediterranean Marine Science, 2(2): 45–65.

Orfanidis, S., Panagiotidis, P. and Stamatis, N. (2003). An insight of the ecological evaluation index (EEI). Ecological Indicators, 3(1): 27–33.

Ormond, R.f.G. and Banimoon, S.A. (1994). Ecology of intertidal macroalgal assemblages on the Hadromount coast of southern Yemen, an area of seasonal upwelling. Marine Ecology Progress Series, 105: 105–120.

Osawa, M., Takahashi, K. and Hay, B.J. (2005). Shell bearing plankton fluxes in the Central Black Sea. Deep Sea Research, I52: 1677–1698.

OSPAR. (2005). Ecological quality objectives for the Greater North Sea with regard to nutrient and eutrophication effects. OSPAR Commission. London.

OSPAR. (2007). Convention on the Protection of the Marine Environment of the North-East Atlantic. OSPAR Commission, London.

OSPAR Agreement. (2010). The North–East Atlantic Environment Strategy: Strategy of the OSPAR Commission for the Protection of the Marine Environment of the North Sea–East Atlantic 2010–2020.

OSPAR. (2017). Eutrophication Status of the OSPAR Maritime Area. Third Integrated Report on the Eutrophication Status of the OSPAR Maritime Area. OSPAR Commission.

Ostman, O., Eklof, J., Eriksson, B.K. et al. (2016). Top-down control as important as nutrient enrichment for eutrophication effects on North Atlantic coastal ecosystems. Journal of Applied Ecology, 53: 1138–1147.

Otto, L. (1983). Currents and water balance in the North Sea. pp. 26–43. *In*: Sundermann, J. and Lenz, W. (Eds.). North Sea Dynamics. Springer, Berlin.

Paerl, H.W., Joyner, J.J., Joyner, A. et al. (2008). Co-occurrence of dinoflagellates and cyanobacterial harmful algal blooms in Florida coastal waters: a case for dual nutrient (N and P) input controls. Marine Ecology Progress Series, 37: 143–153.

Pagou, K. and Gotsis-Skretas, O. (1990). A comparative study of phytoplankton in South Aegean, Levantine and Ionian Seas during March–April 1986. Thalassographica, 13: 13–18.

Pagou, P. and Ignatiades, L. (1988). Phytoplankton seasonality patterns in eutrophic marine coastal waters. Biological Oceanography, 5: 229–241.

Pang, C., Radomyski, A., Subramanian, V., Nadimi-Goki, M., Marcomini, A. and Linkkov, I. (2017). Multi-criteria decision analysis applied to harmful algal bloom management: A case study. Integrated Environmental Assessment and Management, 13(4): 631–639.

Parlamento Italiano. (1999). Decreto Legislativo 11 maggio 1999, n. 152, Allegato I. Monitoraggio e Classificazione delle Aque in Funzione degli Obiettivi di Qualita Ambientale (www. parlamento.it).

Parsons, T.R., Maita, Y. and Lalli, C.M. (1989). A Manual of Chemical and Biological Methods for Seawater Analysis. Pergamon Press, Oxford. 173pp.

Paterson, R.F., McNeil, S., Mitchell, E. et al. (2017). Environmental control of harmful dinoflagellates and diatoms in a fjordic system. Harmful Algae, 69: 1–17.

Patriquin, D.G. and Knowles, R. (1972). Nitrogen fixation in the rhizosphere of marine angiosperms. Marine Biology, 16: 49–58.

Pearman, J.K., Afandi, F., Hong, P. and Carcalho, S. (2018). Plankton community assessment in anthropogenic-impacted oligotrophic coastal regions. Environmental Science and Pollution Research, 25: 31017–31030.

Pearson, T.H. and Rosenberg, R. (1978). Macrobenthic succession in relation to organic enrichment and pollution of the marine environment. Oceanography and Marine Biology, 16: 229–311.

Peirce, B. (1852). Criterion for the rejection of doubtful observations. Astronomical Journal, 2: 161–163.

Peperzak, L. (2003). Climate change and harmful algal blooms in the North Sea. Acta Oceanologica, 24: 139–144.

PERSGA. (2006). State of the Marine Environment, Report for the Red Sea and the Gulf of Aden, PERSGA, Heddah.

Peters, F. and Gross, T. (1994). Increased grazing rates in response to small scale turbulence. Marine Ecology Progress Series, 115: 299–307.

Petersen, J.E., Chen, C.C. and Kemp, W.M. (1997). Scaling aquatic primary productivity: experiments under nutrient and light limited conditions. Ecology, 78: 2326–2338.

Petersen, J.E., Kennedy, V.S., Dennison, W.C. and Kemp, W.M. (Eds.). (2009). Enclosed Experimental Ecosystems and Scale: Tools for Understanding and Managing Coastal Ecosystems. Springer. Berlin. 221pp.

Pettine, M., Casentini, B., Fazi, S., Giovanardi, F. and Pagnotta, R. (2007). A revisitation of TRIX for trophic status assessment in the light of the European Water Framework Directive: application to Italian coastal waters. Marine Pollution Bulletin, 54: 1413–1426.

Pianka, E.R. (1981). Competition and niche theory. pp. 167–196. *In*: May, R.M. (Ed.). Theoretical Ecology: Principles and Applications. Oxford, Blackwell.

Pielou, E.C. (1984). The Interpretation of Ecological Data: A Primer on Classification and Ordination. J. Wiley and Sons, New York.

POEM Group. (1992). General circulation of the Eastern Mediterranean. Earth Science Review, 32: 285–309.

Poole, R.W. (1974). An Introduction to Quantitative Biology. McGraw-Hill, New York.

Portman, M.E. (2016). Environmental Planning for Oceans and Coasts: Methods, Tools and Technologies. Springer. 237pp.

Primpas, I., Karydis, M. and Tsirtsis, G. (2008). Assessment of clustering algorithms in discriminating eutrophic levels in coastal waters. Global Nest Journal, 10: 359–365.

Primpas, I., Tsirtsis, G., Karydis, M. and Kokkoris, G. (2010b). Principal Component Analysis: development of a multivariate index for assessing eutrophication according to the European Water Framework Directive. Ecological Indicators, 10: 178–183.

Primpas, I. and Karydis, M. (2010a). Improving statistical distinctness in assessing trophic levels: the development of simulated normal distributions. Environmental Monitoring and Assessment, 169: 353–365.

Primpas, I. and Karydis, M. (2011). Scaling the trophic index (TRIX) in oligotrophic marine environments. Environmental Monitoring and Assessment, 178: 257–269.

Psarra, S., Tselepides, A. and Ignatiades, L. (2000). Primary productivity in the oligotrophic Cretan Sea (NE Mediterranean): seasonal and international variability. Progress in Oceanography, 46: 187–204.

Raffaelli, D., Raven, J. and Poole, L. (1998). Ecological impact of green macroalgal blooms. Oceanography and Marine Biology. An Annual Review, 36: 97–125.

Rast, W., Jones, R.A. and Lee, G.F. (1983). Predictive capability of U.S. OECD phosphorus loading-eutrophication response models. Journal WPCF, 55(7): 990–1003.

Raven, J.A. and Geider, R.J. (1988). Temperature and algal growth. New Phytologist, 110: 441–461.

Raven, J.A. and Kubler, J.E. (2002). New light on the scaling of metabolic rate with the size of algae. Journal of Phycology, 3: 1–6.

Reichardt, W., McGlone, M.L.S. and Jacinto, G.S. (2006). Organic pollution and its impact on the microbiology of coastal marine environments: a Philippine perspective. Asian Journal of Water Environmental Pollution, 4: 1–19.

Reid, P.C., Taylor, A.H. and Stephens, J.A. (1988). The hydrography and hydrographic balances in the North Sea. *In*: Salomons, W., Bayne, B.L., Duursma, E.K. and Forstrner, U. (Eds.). Pollution of the North Sea: An Assessment. Springer-Verlag.

Reizopoulou, S. and Nikolaidou, A. (2004). Benthic diversity of coastal brackish-water lagoons in western Greece. Aquatic Conservation of Marine and Freshwater Ecosystems, 14(S1): S93–S102.

Relevante, N. and Gilmartin, M. (1995). The relative increase of larger phytoplankton in a subsurface chlorophyll maximum of the Northern Adriatic Sea. Journal of Plankton Research, 17: 1535–1562.

Reynolds, R. (1993). Physical oceanography of the Gulf, Strait of Hormuz and the Gulf of Oman: Results from the Mt. Mitchell expedition. Marine Pollution Bulletin, 27: 35–59.

Rhee, G.Y. (1978). Effects on N:P atomic ratios and nitrate limitation on algal growth, cell composition and nitrate uptake. Limnology and Ocenography, 23: 10–25.

Rhodes, C., Bingham, A. and Heard, A.M. (2017). Diatoms to human uses: linking nitrogen deposition, aquatic eutrophication and ecosystem services. Ecosphere, 8(7): Article e01858.

Richardson, K. and Jorgensen, B.B. (1996). Eutrophication: definition, history and effects. pp. 1–19. *In*: Jorgensen, B.B. and Richardson, K. (Eds.). Eutrophication in Coastal Marine Ecosystems. American Geophysical Union, Washington, D.C.

Romesburg, H.C. (2004). Cluster Analysis for Researchers. Lulu Press. North Carolina.

ROPME. (1988). Marine Monitoring and Research Programme in ROPME Region: Assessment of the Oceanography of the ROPME Sea Area (Summary Report by Cruzado), 27pp.

ROPME. (2013). State of the Marine Environment. Report (SOMER).

Rosas-Luis, R. (2016). Description of plastic remains found in the stomach contents of the jumbo squid Disidicus gigas landed in Ecuador during 2014. Marine Pollution Bulletin, 113: 302–305.

Rosenberg, R. (1995). Eutrophication—the future marine coastal nuisance? Marine Pollution Bulletin, 16(6): 227–231.

Ross, A.D. (1977). The Black Sea and the Sea of Azov. *In*: Nairn, A.E.M. (Ed.). The Ocean Basins and Margins, Vol. 4A, The Eastern Mediterranean, Plenum Press, New York.

Ruhl, J.B., Kraft, S.E. and Lant, C.L. (2007). The Law and Policy of Ecosystem Services. Island Press, Washington.

Ryther, J.H., Hall, J.R., Pease, A.K., Bakun, A. and Jones, M.M. (1966). Primary organic production in relation to the chemistry and hydrography of the Western Indian Ocean. Limnology and Oceanography, 11: 371–380.

Ryther, J.H. and Dunstan, W.M. (1971). Nitrogen, phosphorus and eutrophication in the coastal marine environment. Science, 171: 1008–1013.

Salas, F., Teixeira, H., Marcos, C., Marques, J.C. and Perez-Ruzafa, A. (1988). Applicability of the trophic index TRIX in two transitional ecosystems: the Mar Menor lagoon (Spain) and the Montego Estuary (Portugal). ICES Journal of Marine Sciences, 65: 1442–1448.

Sale, P.F., Feary, D.A. and Burt, J.A. (2011). The growing need for sustainable ecological management of marine communities of the Persian Gulf. Ambio, 40: 4–17.

Salomons, W., Bayne, B.L., Duursma, E.K. and Forstner, U. (1988). Pollution of the North Sea: An Assessment. Springer-Verlag, Berlin. 687pp.

Sands, P., Peel, J., Fabra, A. and MacKenzie, R. (2012). Principles of International Environmental Law. Third edition. Cambridge University Press. 926pp.

Santo, M.L.S., Muniz, K., Barros-Neto, B. and Aranjo, M. (2008). Nutrient and phytoplankton biomass in the Amazon River shelf waters. Annals of the Brazilian Academy of Sciences, 80: 703–717.

Scheffer, M., Szabó, S., Gragnani, A., van Nes, E.H., Rinaldi, S., Kautsky, N., Norberg, J., Roijackers, R.M.M. and Franken, R.J.M. (2003). Floating plant dominance as a stable state. PNAS, 100(7): 4040–4045.

Scheren, P.A., Ibe, A.C., Janssen, F.J. and Lemmens, A.M. (2002). Environmental pollution in the Gulf of Guinea—a regional approach. Marine Pollution Bulletin, 44: 633–644.

Scheren, P.A.G.M. and Ibe, A.C. (2002). Environmental pollution in the gulf of guinea, a regional approach. pp. 299–322. *In*: McGlade, J.M., Cury, P., Koranteng, K.A. and Hardman-Mountford, N.J. (Eds.). The Gulf of Guinea: Large Marine Ecosystem: Environmental Forcing and Sustainable Development of Marine Resources. Elsevier. Amsterdam.

Schnetzer, A., Lampe, R.H., Benitez-Nelson, C.R. et al. (2017). Marine snow formation by the toxin-producing diatom, Pseudonitzschia australis. Harmul Algae, 61: 23–30.

Schott, F., Xie, S.P. and McGreary, J.P. (2009). Indian Ocean: circulation and climatic variability. Rev. Geophys. 47: 1–46.

Schulman, P.R. (1975). Nonicremental Policy Making: Notes Toward and Alternative Paradigm. American Political Science Review, Volume 69, No. 4.

Schuwirth, N, Honti, M., Logar, I. and Stamm, C. (2018). Multi-criteria decision analysis for integrated water quality assessment and management support. Water Research X, 1: 100010.

Segar, D.A. and Stamman, E. (1986). Fundamentals of marine pollution monitoring program design. Marine Pollution Bulletin, 17(5): 194–200.

Segar, D.A. (1997). Introduction to ocean sciences. Wadsworth Publ. Co. Belmont, CA. 497pp.

Seitzinger, S.P., Harrison, J.A., Dumont, E. et al. (2005). Sources and delivery of carbon, nitrogen and phosphorus to the coastal zone: an overview of global nutrient export from watersheds (NEWS) models and their applications. Global Biogeochemical Cycles, 19: GB4S01.

Semerci, A., Kaya, Y. and Durak, S. (2007). Economic analysis of sunflower production in Turkey. Helia, 30: 105–114.

Sfiso, A., Pavoni, B., Marcomini, A. and Orio, A.A. (1988). Annual variations of nutrients in the Lagoon of Venice. Marine Pollution Bulletin, 19: 54–60.

Shahrban, M. and Etemad-Shahidi, A. (2010). Classification of the Caspian Sea coastal waters based on trophic index and numerical analysis. Environmental Monitoring and Assessment, 164: 349–356.

Sharma, S. (1996). Applied Multivariate Techniques. J. Wiley and Sons, Inc.

Sheehy, B. (2004). Does International Marine Environmental Law Work? An examination of the cartagena convention for the Wider Caribbean Region. Georgetown International Environmental Law Review, 12: 441–472.

Sheppard, C., Price, A. and Roberts, C. (1992). Marine Ecology of the Arabian Region. Academic Press, London.

Sheppard, C. (1993). Physical environment in the Gulf relevant to marine pollution: an overview. Marine Pollution Bulletin, 27: 3–8.

Sheppard, C., Al-Husiani, M., Al-Jamali, F. et al. (2010). The Gulf: a young sea in decline. Marine Pollution Bulletin, 60: 13–38.

Sherman, K. and Anderson, E.D. (2002). A modular approach to monitoring, assessing and managing large marine ecosystems. pp. 9–25. *In*: McGlade, J.M., Cury, P., Koranteng, K.A. and Hardman-Mountford, N.J. (Eds.). The Gulf of Guinea: Large Marine Ecosystem: Environmental Forcing and Sustainable Development of Marine Resources. Elsevier. Amsterdam.

Short, F., Carruthers, T., Dennison, W. and Waycott, M. (2007). Global seagrass distribution and diversity: a biological model. Journal of Experimental Marine Biology and Ecology, 350: 3–20.

Shriadah, M. and Al-Gais, S. (1999). Environmental characteristics of the United Arab Emirates waters along the Arabian Gulf Hydrographical survey and nutrient slats. Indian Journal of Marine Sciences, 28: 225–232.

Shucksmith, R., Gray, L., Kelly, C. and Tweddle, J.F. (2014). Regional Marine Spatial Planning: the data collection and mapping process. Marine Policy, 50: 1–9.

Silkin, V.A., Pantova, L.A., Pakhomova, S.V., Lifanchuk, A.V., Yakushev, E.V. and Chasovnikov, V.K. (2014). Environmental control on phytoplankton community structure in the NE Black Sea. Journal of Experimental Marine Biology and Ecology, 461: 267–274.

Simboura, N. and Zenetos, A. (2002). Benthic indicators to use in ecological quality classification of Mediterranean soft bottom marine ecosystems, including a new biotic index. Mediterranean Marine Science, 3/2: 77–111.

Simboura, N., Panayotidis, P. and Papathanassiou, E. (2005). A synthesis of the biological quality elements for the implementation of the European Water Framework Directive in the Mediterranean ecoregion: the case of Saronikos Gulf. Ecol. Indic, 5(3): 253–266.

Simboura, N., Pavlidou, A., Bald, J., Tsapakis, M., Pagou, K., Zeri, Ch., Androni, A. and Panayotidis, P. (2016). Response of ecological indices to nutrient and chemical contaminant stress factors in Eastern Mediterranean coastal waters. Ecological Indicators, 70: 89–105.

Siokou-Frangou, I., Christaki, U., Mazzocchi, M.G., Montresor, M., Ribera d' Alcala, Vaque, M. and Zingone, A. (2010). Plankton in the open Mediterranean Sea: a review. Biogeosciences Discussion, 6: 11187–11293.

Siung-Chang, A. (1997). A review of marine pollution issues in the Caribbean. Environmental Geochemistry and Health, 19: 45–55.

Sling, J., Spencer, H., Hosking, B., Berrisford, P. and Black, E. (2005). The meteorology of the West Indian Ocean and the influence of the East African Highlands. Phil. Trans. R. Soc. A363 (1826): 25–42.

Small, C. and Nicholls, R.J. (2003). A global analysis of human settlement in coastal zones. Journal of Coastal Research, 19(3): 584–589.

Smayda, T. (1990). Novel and nuisance phytoplankton blooms in the sea: evidence for a global epidemic. pp. 29–40. *In*: Graneli, E., Sundstorm, B. and Edler, L. (Eds.). The Toxic Marine Phytoplankton. Elsevier. New York.

Smith, S.V., Swaney, D.P., Talue-Mcmanus, L. et al. (2003). Humans, hydrology and the distribution of inorganic nutrient loading to the Ocean. BioScience, 53: 235–245.

Smith, V.H., Tilman, G.D. and Nekkola, J.C. (1999). Eutrophication: impacts of excess nutrient inputs on freshwater, marine and terrestrial ecosystems. Environmental Pollution, 100: 179–196.

Sokal, R.R. and Rohlf, F.J. (1981). Biometry. Freeman. New York.

Sommer, U. (1988). Phytoplankton succession in microcosm experiments under simultaneous grazing pressure and resource limitation. Limnology and Oceanography, 33(5): 1037–1054.

Sommer, U. (1989). Nutrient status and nutrient competition of phytoplankton in a shallow hypertrophic lake. Limnology and Oceanography, 34(7): 1162–1173.

Sommer, U. (1991). Convergent succession of phytoplankton in microcosms with different inoculum species composition. Oecologia, 87: 171–179.

Sonnemann, G. and Valdivia, S. (2017). Medellin declaration on marine litter in life cycle assessment and management. International Journal of Life Cycle Assessment, 22: 1637–1639.

Sournia, A., Chretiennot-Dinet, M.J. and Richard, M. (1991). Marine phytoplankton: how many species in the world ocean? Journal of Plankton Research, 13: 1093–1099.

Souvermezoglou, E., Krasakopoulou, E. and Pavlidou, A. (1999). Temporal variability in oxygen and nutrient concentrations in the southern Aegean Sea and the Straits of the Cretan Arc. Progress in Oceanography, 44: 573–600.

Spalding, M., Blasco, F. and Field, C.D. (Eds.). (1997). World Mangrove Atlas. The International Society for Mangrove Ecosystems. Okinawa, Japan.

Spatharis, S. and Tsirtsis, G. (2010). Ecological quality scales based on phytoplankton for the implementation of Water Framework Directive in the Eastern Mediterranean. Ecological Indicators, 10: 840–847.

Spatharis, S., Roelke, D.L., Dimitrakopoulos, P.G. and Kokkoris, G.D. (2011). Analyzing the (mis) behavior of Shannon index in eutrophication studies using field and simulated phytoplankton assemblages. Ecological Indicators, 697–703.

Spies, R.B. (2011). The biological effects of petroleum hydrocarbons in the sea: assessment from the field and microcosms. pp. 411–468. *In*: Boesch, D.F. and Rabalais, N.N. (Eds.). Long-Term Environmental Effects of Offshore Oil and Gas Development. Taylor and Francis. London and New York.

Spitieri, E.D. (2017). Assessment of Water Quality in a Coastal System. Thesis, Univ. of the Aegean, Dept. of Marine Sciences (in Greek with Abstract in English).

SPREP. (1981). Marine Pollution in the South Pacific (by C.A. Matos). Topic Review No. 11. South Pacific Commission, Noumea, New Caledonia.

SPREP. (2010). Report of the Tenth Meeting of the Conference of the parties to the Noumea Convention, 2 September 2010, Madang, Papua New Guinea – Apia Samoa, 46pp.

Stambler, N. (2005). Bio-optical properties of the Northern Red Sea and the Gulf of Eilat (Aqaba) during winter 1999. Journal of Sea Research, 54: 186–203.

Steele, J.H. (1974). The Structure of Marine Ecosystems. Harvard University Press, Cambridge MA.

Stefanou, P., Tsirtsis, G. and Karydis, M. (2000). Nutrient scaling for assessing eutrophication: the development of a simulated normal distribution. Ecological Applications, 10: 303–309.

Stevens, J.P. (1996). Applied Multivariate Statistics for the Social Sciences. Lawrence Erbaum Associates, Inc., Publishers. New Jersey.

Stolte, W. and Granelli, E. (2006). Threshold nutrient levels for harmful algal events in some European coastal waters. THRESHOLDS: Thresholds of Environmental Sustainability, D3.1.2, Sixth Framework Programme; http://www.thresholds-eu.org/.

Strokal, M. and Kroeze, C. (2013). Nitrogen and phosphorus inputs to the Black Sea in 1970–2050. Reg. Environ. Chang. 13(1): 179–192.

Sundermann, J. (Ed.). (1994). Circulation and Contaminant Fluxes in the North Sea. Springer-Verlag.

Sur, H.L., Ozsoy, E., Ilyin, Y.P. and Unluata, U. (1996). Coastal/deep ocean interactions in the Black Sea and their ecological/environmental impacts. Journal of Marine Systems, 7: 293–320.

Swan, S.C., Turner, A.D., Bresnan, E. et al. (2018). Dinophysis acuta in Scottish Coastal Waters and its influence on diarrhetic shellfish toxin profiles. Toxins, 10: 399.

Sylaios, G., Koutrakis, E. and Kallianiotis, A. (2006). Hydrographic variability, nutrient distribution and water mass dynamics in Strymonikos Gulf (Northern Greece). Continental Shelf Research, 26: 217–235.

Taebi, S., Etemad-Sahidi, A. and Fardi, G.A. (2005). Examination of three eutrophication indices to characterize water quality in the north east of Persian Gulf. Journal of Coastal Research, 42: 405–411.

Taketani, F., Aita, M.N., Yamaji, K. et al. (2018). Seasonal response of North Western Pacific marine ecosystems to deposition of atmospheric inorganic nitrogen compounds from East Asia. Scientific Reports, 8: 9324–9333.

Tarrason, L., Olendrzynski, K., Stiren, M.P., Bartnicki, J. and Vestreng, V. (2000). Transboundary acidification and eutrophication in Europe. *In*: Tarasson, L. and Schaug, J. (Eds.). EMEP Report 1/100.

Taslakian, M.J. and Hardy, J.T. (1976). Sewage nutrient enrichment and phytoplankton ecology along the central coast of Lebanon. Marine Biology, 38: 315–325.

Taylor, A.H., Reid, P.C., Marsh, T.J., Stephens, J.A. and Jonas, T.D. (1981). Year-to-year changes in the salinity of the Eastern English Channel 1948–1973: a budget. Journal of the Marine Biological Association of United Kingdom, 61: 489–450.

TDA. (2007). Black Sea Transboundary Diagnostic Analysis. Available at: http://www.blacksea-commission.org.

Teen, L.P., Gires, U. and Pin, L.C. (2012). Harmful Algal Blooms in Malay Sian Waters. Sains Malysiana, 41: 1509–1515.

Teng, S.K. (2006). Practitioner Guidelines for Preparation of Transboundary Diagnostic Analysis (TDA) and Strategic Action Programme (SAP) in East Asian Seas Region. UNEP: Southeast Asia Regional Learning Centre, Bangkok, Thailand.

Thieu, V., Garnier, J. and Billen, G. (2010). Assessing the effect of nutrient mitigation measures in the watersheds of the Southern Bight of the North Sea. Science of the Total Environment, 408: 1245–1255.

Tomczak, M. and Godfrey, J.S. (1994). Regional Oceanography: An Introduction. Pergamon Press.

Torbick, N., Hu, F., Zhang, J., Qi, J., Zhang, H. and Becker, B. (2008). Mapping chlorophyll-a concentrations in West Lake, China using Landsat 7 ETM+. Journal of Great Lakes Research, 34(3): 559–565.

Tsirtsis, G. and Karydis, M. (1998). Evaluation of phytoplankton community indices for detecting eutrophic trends in the marine environment. Environmental Monitoring and Assessment, 50: 255–269.

Tsirtsis, G. and Karydis, M. (1999). Application of discriminant analysis for water quality assessment in the Aegean. 6th International Conference of Environmental Science and Technology. Samos, Greece.

Tsirtsis, G. and Karydis, M. (1996). Aquatic microcosms: a methodological approach for the quantification of eutrophication processes. Environmental Monitoring and Assessment, 48: 193–215.

Turner, R.K. et al. (1999). Managing nutrient fluxes and pollution in the Baltic: an interdisciplinary simulation study. Marine Ecology Progress Series, 172: 13–24.

Turner, R.K. and Bateman, I.J. (Eds.). (2001). Water Resources and Coastal Management. Edward Elgar Publishers. Northampton MA. 527pp.

Turner, R.K. and Schaafsma, M. (Eds.). (2015). Coastal Zone Ecosystem Services: From Science to Values and Decision Making. Springer. 240pp.

Tuzhilin, V.S. and Kosarev, A.N. (2005). The thermohaline structure and general circulation of the Caspian Sea waters. pp. 33–57. In: Kostianoy, A. and Kosarev, A. (Eds.). The Caspian Sea Environment. Springer.

Tyrrell, T., Maranon, E., Poulton, A., Bowie, R., Harbour, S. and Woodward, E.M.S. (2003). Large-scale latitudinal distribution of Trichodesmium spp., in the Atlantic Ocean. Journal of Plankton Research, 25: 405–416.

Uchida, H., Watanabe, R., Matsushima, R. et al. (2018). Toxin profiles of Okadaic Acid Analogus and Other Lipophilic Toxins in Dinophysis from Japanese Coastal Waters. Toxins, 10: 457.

Uitz, J., Clauste, H., Gentili, B. and Stramski, D. (2010). Phytoplankton class-specific primary production in the world's oceans. Seasonal and interannual variability form satellite observations. Global Biogeochemical Cycles, 24: GB3016.

UN OECD. (2016). Environmental Performance Reviews. Peru. Highlights and Recommendations. A United Nations—OECD Publication.

Underwood, A.J. (1981). Techniques of analyses of variance in experimental marine biology and ecology. Oceanography and Marine Biology Annual Reviews, 19: 513–605.

Underwood, A.J. (2007). Experiments in Ecology. Cambridge University Press. Cambridge.

UNEP. (1989). State of the Mediterranean Marine Environment. MAP Technical Series No. 28, UNEP, Athens.

UNEP. (1999). Overview of Land-Base Sources and Activities Affecting the Marine, Coastal and Associated Freshwater Environments in the West and Central Africa Region. UNEP Regional Seas Reports and Studies No. 171.

UNEP. (2000). Overview on Land-Based Sources and Activities Affecting the Marine Environment in the East Asian Seas. United Nations, 74pp.

UNEP. (2002). African Environmental Outlook—Past, Present and Future Perspectives. Web Publication: Grid-Arendal: http://www.grida.no/publications/other/aeo/?src=/aeo/.

UNEP. (2003a). Eutrophication monitoring and strategy of MED POL. UNEP(DEC)/MED WG.231/14.

UNEP. (2003b). Assessment of transboundary pollution issues in the Mediterranean Sea. MAP Report, UNEP(DEC)/MED, W.G. 228/Inf.7.

UNEP. (2005). East Asian seas region. UNEP Regional Seas Reports and Studies, 1–34.

UNEP. (2006a). Challenges to International Waters—Regional Assessments in a Global Perspective. United Nations Environment Programme, Nairobi, Kenya.

UNEP. (2006b). The State of the Marine Environment. Regional Assessment. The Hague, July 2006.

UNEP. (2008). New Strategic Direction for COBSEA (2008–2012). COBSEA Secreteriat, United Nations Environment Programme, 23.

UNEP. (2009). State of the Environment and Development in the Mediterranean Plan Bleu, UNEP/MAP's Regional Agency Centre.

UNEP. (2010). Global Synthesis: A Report from the Regional Seas Conventions and Action Plans for the Marine Biodiversity Assessment and Outlook. Nairobi, United Nations Environment Programme.

UNEP/EAF. (2012). Decisions, UNEP(DEPI)/EAF/CP. 7/6b-En EP: Draft Guidelines for drafters and negotiators of the ICZM Protocol to The Nairobi Convention. Presented at the seventh conference of parties to the Convention of the Protection, Management and development of the marine and coastal environment of the Western Indian Ocean. Maputo, Mozambique 10–14 December, 2012.

UNEP. (2013). Africa Environment Outlook 3: Our Environment, Our Health, Summary for Policy Makers. 40pp.

UNEP. (2015). Ecosystem Approaches to Regional Seas.

UNEP/COBSEA. (2010). State of the Marine Environment Report for the East Asian Sea. 2009 (Ed. Chou, L.M.). COBSEA Secreteriat, Bangkok, 156p.

UNEP/COBSEA. (2018). Second Extraordinary Intergovernmental Meeting of the Coordinating Body on the Seas of East Asia (COBSEA), Bangkok, Thailand 25–26 April 2018. UNEP/OBSEA IGM EO 2/6.

UNEP/FAO/WHO. (1996). Assessment of the State of Eutrophication in the Mediterranean Sea. MAP Technical Series No. 106, UNEP, Athens.

UNEP/GEF/Kalmar Högskola/Cimab. (2004). Global International Water Assessment (GIWA), Caribbean Islands Bahamas, Cuba, Dominican Republic, Haiti, Jamaica, Puerto Rico Regional Assessment 4, Kalmar Sweden.

UNEP/GEF/Kalmar Högskola, Invemar. (2006). Global International Water Assessment (GIWA), Caribbean Sea/Colombia & Venezuela, Central America & Mexico GIWA Regional Assessment 3b, 3c, Kalmar Sweden.

UNEP/MAP. (2007). Eutrophication Monitoring Strategy for the MED POL (REVISION). UNEP(DEPI)/MED WG.321/Inf. 5, 9 November 2007, Athens.

UNEP/MAP-Plan Bleu. (2009). State of the Environment and Development in the Mediterranean, UNEP/MAP-Plan Bleu, Athens.

UNEP–Nairobi Convention Secretariat. (2009). Strategic Action Programme of the Protection of the Coastal and Marine Environment of the Western Indian Ocean from Land-Based Sources and Activities. Nairobi, Kenya.

UNEP–Nairobi Convention, CSIR and WIOMSA. (2009a). Regional Synthesis Report on the Status of Pollution in the Western Indian Ocean Region. UNEP. Nairobi, Kenya.

UNEP–Nairobi Convention Secretariat and WIOMSA. (2009b). Transboundary Diagnostic Analysis of Land-Based Sources and Activities Affecting the Western Indian Ocean Coastal And Marine Environment. UNEP. Nairobi, Kenya.

UNEP–Nairobi Convention Secretariat. (2009c). Regional Review of Policy, Legal and Institutional Frameworks for Addressing Land-Based Sources and Activities in the Western Indian Ocean Region. UNEP. Nairobi, Kenya.

UNEP–Nairobi Convention and WIOMSA. (2015). The Regional State of the Coast Report: Western Indian Ocean. UNEP and WIOMSA, Nairobi, Kenya, 546pp.

UNEP(OCA)/CAR. (1994). Regional Overview of Land-Based Sources of Pollution in the Wider Caribbean Region (Revised Version WG. 14/4, 55), UNEP(OCA)/CAR.

UNEP/SCS. (2008). Strategic Action Programme for the South China Sea. UNEP/GEF/SCS, Technical Publications No. 16.

UNESCO. (2013). Marine Spatial Planning Initiative, Marine Spatial Planning. http://www.un.org/sg/staements/index.asp?nid=6239, (accessed 16.09.2016).

United Nations. (2017). The First Global Integrated Marine Assessment I. Cambridge University Press. 973pp.

Vadrucci, M.R., Sabetta, L., Fioca, A. et al. (2008). Statistical evaluation of differences in phytoplankton richness and abundance as constrained by environmental drivers in transitional waters of the Mediterranean basin. Aquatic Conservation of Marine and Freshwater Ecosystems, 18: S88–S104.

Valiela, I. (1995). Marine Ecological Processes. New York: Springer-Verlag.

Van Bennekom, A.J., Gieskes, W.W.C. and Tijssen, S.B. (1975). Eutrophication of Dutch coastal waters. Proceedings of the Royal Society of London < B, 189: 359–374.

Van der Belt, M. et al. (2017). Scientific understanding of ecosystem services. pp. 67–89. *In*: United Nations (Ed.). The First Global Integrated Marine Assessment I. Cambridge University Press. 973pp.

Van Dykem, J.M. and Broder, S.P. (2017). International assessments and customary international principles providing guidance for National and Regional Ocean Policies. pp. 49–84. *In*: Cicin-Sain, B., VanderZwaag, D. and Balgos, M.C. (Eds.). Routledge Handbook of National and Regional Ocean Policies. Routledge, Taylor and Francis Group. London.

Van Holt, T., Moreno, G.A., Binford, M.W., Portier, K.M., Mulsav, S. and Frazer, T. (2012). Influence of landscape change on nearshore fisheries in Southern Chile. Global Change Biology, 18: 2147–2160.

Van Lavieren, H. and Klaus, R. (2013). An effective regional marine protected area network for the ROPME sea area; unrealistic vision or realistic possibility? Marine Pollution Bulletin, 72: 389–405.

van Tongeren, O.F.R. (1987). Cluster analysis. pp. 174–212. *In*: Jongman, R.H.G., ter Braak, J.F. and van Tongeren, F.R. (Eds.). Data Analysis in Community and Landscape Ecology. Pudoc Wageningen. Netherlands. 299pp.

Vaque, D., Marrase, C., Iniguez, V. and Alcaraz, M. (1989). Zooplankton influence on the phytoplankton–bacterioplankton coupling. Journal of Plankton Research, 11(3): 625–632.

Viaroli, P., Bartoli, M., Giordani, G. et al. (2008). Community shifts, alternative stable states, biochemical controls and feedback in eutrophic coastal lagoons: a brief overview. Aquatic Conservation: Marine and Freshwater Ecosystems, 18: S105–S117.

Vascetta, M., Kauppila, P. and Furman, E. (2008). Aggregate indicators in coastal policy making: potentials of the trophic index TRIX for sustainable considerations of eutrophication. Sustainable Development, 16: 282–289.

Vassiliou, A., Ignatiades, L. and Karydis, M. (1989). Clustering of transect phytoplankton collections with a quick randomization algorithm. Journal of Experimental Marine Biology and Ecology, 130: 135–145.

Vega, M., Pardo, R., Barrado, E. and Deban, L. (1998). Assessment of seasonal and polluting effects on the quality of river water by exploratory data analysis. Water Research, 32: 3581–3592. EMEP/MSC-W Note 1/2002.

Verlaan, P.A. (1991). South African Seas. Occasional Reports of East-west Environment and Policy Institute. Papers No. 14, East-West Center, Honolulu.

Vespremeanu, E. and Golumbeanu, M. (2018). The Black Sea; Physical, Environmental and Historical Perspectives. Springer.

Vestreng, V. and Klein, H. (2002). Emission Data Report to UNECE/EMEP: Quality Assurance, Trend Analysis and Presentation of WebDab. MSC-W Status Report 2002.

Vilicic, D., Silovic, T., Kuzmic, M. et al. (2011). Phytoplankton distribution across the southeast Adriatic continental and Shelf slope to the west of Albania (spring aspect). Environmental Monitoring and Assessment, 177: 593–607.

Vollenweider, R.A. (1969). Possibilities and limits of elementary models concerning the budgets of substances in lakes. Arch. Hydrobiol., 66: 1–36.

Vollenweider, R.A. (1974). A Manual of Methods for Measuring Primary Productivity in Aquatic Environments. 2nd Edition. Blackwell Scientific Publications, Oxford. 225pp.

Vollenweider, R.A. (1976). Advances in defining critical loading levels for phosphorus in Lake eutrophication. Memorie dell' Istituto Italiano di Idrobiologia, 33: 53–83.

Vollenweider, R.A. (1976). Advances in defining critical loading levels of phosphorus in lake eutrophication. Mem Ist Ital Idrobiol, 33: 53–83.

Vollenweider, R.A. (1992). Coastal marine eutrophication: principles and control. pp. 1–20. In: Vollenweider, R.A., Marchetti, R. and Viviani, R. (Eds.). Marine Coastal Eutrophication. Elsevier, London.

Vollenweider, R.A., Giovanardi, F., Montanari, G. and Rinaldi, A. (1998). Characterization of the trophic conditions of marine coastal waters with special reference to the NW Adriatic Sea: proposal for a trophic scale, turbidity and generalized water quality index. Environmetrics, 9: 329–357.

Voogd, H. (1983). Multicriteria Evaluation for Urban and Regional Planning. London: Pion.

Vounatsou, P. and Karydis, M. (1991). Environmental characteristics in oligotrophic waters: data evaluation and statistical limitations in water-quality studies. Environmental Monitoring and Assessment, 18: 211–220.

Vukadin, I. (1992). Impact of nutrient enrichment and its relationship to the algal bloom of Adriatic Sea. pp. 364–369. In: Vollenweider, R.A., Marchetti, R. and Viviani, R. (Eds.). Marine Coastal Eutrophication. Proceedings of an International Conference, Bologna, Italy, 21024 March 1990. Science of the Total Environment, Elsevier.

Wafar, M., Venkatamaran, K., Ingole, B., Khan, S.A. and Loka Bhazathi, P. (2011). State of knowledge of coastal and marine biodiversity of Indian Ocean Countries. Plos ONE, 6(1).

Wagner, C. and Adrian, R. (2009). Cyanobacteria dominance: quantifying the effects of climate change. Limnology and Ocenography, 54: 2460–2468.

Walsby, A.E., Hayes, P.K., Boje, R. and Stal, L.J. (1977). The selective advantage of buoyancy provided by gas vesicles for planktonic cyanobacteria in the Baltic Sea. New Phytologist, 136: 407–417.

Walz, R. (2000). Development of environmental indicator species: experiences from Germany. Environmental Management, 25(6): 613–623.

Wang, J.J., Jing, Y.Y., Zhang, C.F. and Zhao, J.H. (2009). Review on multi-criteria decision analysis aid in sustainable energy decision-making. Renewable & Sustainable Energy Reviews, 13: 2263–2278.

Wang, S.F., Tang, D.L., He, F.L., Fukuyo, Y. and Azanza, R.V. (2008). Occurrences of harmful algal blooms (HABs) associated with ocean environments in the South China Sea. Hydrobiologia, 596: 79–93.

Warwick, R.M., Carr, M.R., Clarke, K.R., Gee, J.M. and Green, R.H. (1988). A mesocosm experiment on the effects of hydrocarbon and copper pollution on a sublittoral soft-sediment meiobenthic community. Marine Ecology Progress Series, 46: 181–191.

Washington, H.G. (1984). Diversity biotic and similarity indices: a review with special relevance to aquatic ecosystems. Water Research, 18: 653–694.

Wasmund, N., Andrushaitis, A., Lysiak-Pastuzak, E. et al. (2001). Primary production rates calculated by different concepts—an opportunity to study the complex production system in the Baltic Proper. Journal of Sea Research, 54: 244–255.

Wasmund, N. and Uhlig, S. (2003). Phytoplankton trends in the Baltic Sea. ICES Journal of Marine Science, 60: 177–186.

Weber, D.D. and Englund, E.J. (1994). Evaluation and comparison of spatial interpolators 2. Mathematical Geology, 26: 589–603.

Webster, R. and Oliver, M.A. (2001). Geostatistics for Environmental Scientists. Australia: John Wiley and Sons.

Werdell, P.J., Bailey, S.W., Franz, B.A., Harding, Jr. L.W., Feldman, G.C. and McClain, C.R. (2009). Regional and seasonal variability of chlorophyll-a in Chesapeake Bay as observed by SeaWiFS and MODIS-Aqua. Remote Sensing of Environment, 113(6): 1319–1330.

Whittaker, R.H. (Ed.). (1982). Ordination of Plant Communities. Dr. W. Junk Publishers. The Hague. 388pp.

WHO/UNEP. (1996). Survey of pollutants from land-based sources in the Mediterranean. MAP Technical Reports Series No. 109, MAP, Athens.

WHO/UNEP. (1997). Identification of priority pollution hot spots and sensitive areas in the Mediterranean. Document UNEP (OCA)/MED IG.11/Inf.8, UNEP, Athens.

Wiens, J.A. (2001). Understanding the problem of scale in experimental ecology. pp. 61–88. *In*: Gardner, R.H., Kemp, W.M., Kennedy, V.s. and Petersen, J.E. (Eds.). Scaling Relations in Experimental Ecology. Columbia University Press, New York. 373pp.

Wilhm, L. and Dorris, C.T. (1968). Biological parameters for water quality criteria. BioScience, 18: 477–481.

Wilson, K.A.A., Able, K.W. and Heck, K.L. (1990). Predation rates on juvenile blue crabs in estuarine nursery habitats: evidence for the importance of macroalgae (*Ulva lactula*). Marine Ecology Progress Series, 58: 243–251.

Winsemius, P. (1986). Gast in eigen huis. Samson/Tjeenk Willink.

Woodard, C. (1997). Black Sea Nations Wary of Russia's Big Oil Plans. Christian Science Monitor, 22 October.

WWF. (2008). Effects of Climate Change on Eutrophication in the Northern Baltic Sea. Available at: www.panda.org/europe/baltic.

Wynne, S.P. (2017). Observational Evidence of Regional Eutrophication in the Caribbean Sea and Potential Impacts on Coral Reef Ecosystems and their Management in Anguilla, BW1, Anguilla Fisheries and Marine Resources, Research Bulletin, No. 8. Available on line at www.gov.ai/documents/fisheries.

Wyrtki, K. (1973). Physical oceanography of the Indian Ocean. pp. 18–36. *In*: Zeitschel, B. (Ed.). The Biology of the Indian Ocean. Springer, Berlin.

Xu, F.L., Tao, S., Dawson, R.W. and Li, B.G. (2001). A GIS-based method of lake eutrophication assessment. Ecological Modeling, 144: 231–244.

Xu, N., Huang, B., Hu, Z. et al. (2017). Effects of temperature, salinity and irradiance on the growth of harmful algal bloom species, Phaeocystis globose Schrfell (Prymnesiophyceae), isolated from the South China Sea. Chinese Journal of Oceanology and Limnology, 35: 557–565.

Yan, X., Huang, H., Zhou, W. et al. (2016). Long term variations in oxygen in sub-tropical coastal waters: influence of sewage effluent. Aquatic Ecosystem Health and Management, 19: 336–344.

Yamaguchi, H., Hirano, T., Yoshimatsou, T. et al. (2016). Occurrence of *Karenia papilionacea* (Dinophyceae) and its novel sister phylotype in Japanese coastal waters. Harmful Algae, 57: 59–68.

Yankovsky, A.E., Lemeshko, E.M. and Ilyin, Y.P. 2004. The influence of shelfbreak forcing on the alongshelf penetration of the Danube buoyant water, black sea. Cont. Shelf Res. 24(10): 1083–1098.

Yunev, O.A., Verdenikov, V.I., Basturk, O., Yilmaz, A., Kideys, A.E., Moncheva, S. and Konovalov, S.K. (2002). Long term variations of surface chlorophyll α and primary production in the open Black Sea. Marine Ecology Progress Series, 230: 11–28.

Yunev, O.A., Carstensen, J., Moncheva, S., Khaliulin, A., Ertebjerg, G. and Nixon, S. (2007). Nutrient and phytoplankton trends on the western Black Sea shelf in response to cultural eutrophication and climate changes. Estuarine, Coastal and Shelf Science, 74: 63–76.

Yunev, O., Velikova, V. and Carstensen, J. (2017). Reconstructing the trophic history of the Black Sea shelf. Continental Shelf Research, 150: 1–9.

Zaitsev, Y. (1994). Impact on Eutrophication on the Black Sea Fauna. General Fisheries Counsel for the Mediterranean. Studies and Reviews, Rome, FAO.

Zaldívar, J.M., Cardoso, A.C., Viaroli, P., Newton, A., de Wit, R., Ibañez, C., Reizopoulou, S., Somma, F., Razinkovas, A., Basset, A., Holmer, M. and Murray, N. (2008). Eutrophication in transitional waters: an overview. Transitional Waters Monographs, 1: 1–78.

Zaldívar, J.M., Bacelar, F.S., Dueri, S., Marinov, D., Viaroli, P. and Hernández-García, E. (2009). Modeling approach to regime shifts of primary production in shallow coastal ecosystems. Ecological Modelling, 220: 3100–3110.

Zenkevich, L. (1963). Biology of the Seas of USSR. Allen and Unwin, London.

Zhang, Z., Rengel, Z. and Meney, K. (2008). Interactive effects of nitrogen and phosphorus loadings on nutrient removal from simulated wastewater using *Schoenoplectus validus* in wetland microcosms. Chemosphere, 72: 1823–1828.

Zonn, I.S. (2005). Economic and international legal dimensions. pp. 243–256. *In*: Kostianoy, A. and Kosarev, A. (Eds.). The Caspian Sea Environment. Springer.

Index